human engineering

human engineering

the body re-examined

John Lenihan

George Braziller, New York

82-474

Published in the United States of America in 1975
by George Braziller, Inc.

Copyright © 1974 by John Lenihan

Originally published in England by Weidenfeld and Nicolson

Text illustrations by John B. Fleming

International Standard Book Number: 0-8076-0782-7
Library of Congress Catalog Card Number: 74-25318
Second Printing, June 1976
Printed in the United States of America

contents

illustrations

Figure 11 is adapted (by courtesy of Time-Life Books) from a diagram which appeared in 'The Body' by Allan E Nourse.

1: the engineer's view of man

The proper study of mankind is man . . .

Alexander Pope was not expressing a new thought. Even in 1733, the study of man had long been a fascinating task for philosophers, physicians and poets. To this day there are universities in Scotland where Latin is taught by professors of humanity — reminding us of the time when knowledge was divided into divinity (the study of God) and humanity, the study of man.

In the biological sense, man might seem to be not very remarkable. As far as we know, most of the universe is uninhabited — and even on the earth, the total mass of all living things, including plants and micro-organisms, is less than a millionth of the mass of the rocks, sea and atmosphere. In numbers and in bulk, man does not predominate. His structure, function, adaptation and biochemical arrangements are interesting — but, to the biologist, not exceptional since most of their distinctive features can be matched or surpassed elsewhere in the living world.

But man is unique in two ways. First, he is the only creature endowed with the introspective ability to reflect on what he is and with the technological resources to modify his internal or external environment in a purposeful way. Secondly he is the only creature to engage in systematic slaughter of his own kind.

This book is concerned with the technology of man. It will be useful to define what we mean by technology — and by science, since the two activities are so often confused. Science is concerned with abstractions, such as forces, atoms and genes. Technology deals with concrete realities, such as materials and machines. In computer language, science is software and technology is hardware.

Technology provides the equipment and materials with which the scientist seeks to verify his speculations and theories, but there is a closer relationship between the two act ities. Science proceeds by asking questions or, what amounts to the same thing, by making models and testing them against reality. The models are not usually tangible structures but are analogies or theories relating the structure, function and behaviour of the system under investigation to the known properties of some other system. The models by which science advances are usually based on currently fashionable technology. Not so long ago, the brain was compared to a telephone exchange; more recently the electronic computer provided the model and, later still, new developments in optics were invoked to explain the brain as a holographic data storage system.

This book aims to provide a new model of man. It is not the first attempt of its kind. In the seventeenth century, when the mechanics of Galileo and Newton seemed capable of solving every problem, it was fashionable to consider the body as an assembly of levers, cords, pulleys and bellows. Richard Mead, who numbered among his patients Queen Anne (and, more significantly, Isaac Newton) went even further. His book *A Mechanical Account of Poisons*, published in 1702, began with the claim that the study of mathematics would show doctors how to solve the intractable problems of medicine. The idea was interesting but its fulfilment defeated Mead, and his book, after a stirring preface, degenerated into a collection of anecdotes and empirical remedies.

Mead's contemporary Giorgio Baglivi, Professor of Anatomy in Rome, took a more mature view of the usefulness of the technologist's model of man. The human body, he wrote, 'as to its natural actions, is truly nothing else but a Complex of Chymico-Mechanical Motions, depending upon such Principles as are purely Mathematical. For whoever takes an attentive View of its Fabrick, he'll really meet with Shears in the Jaw-bones and Teeth . . . Hydraulick Tubes in the Veins, Arteries and other Vessels . . . a Pair of Bellows in the Lungs, the Power of a Leaver in the Muscles, Pulleys in the Corners of the Eyes, and so on.'

Baglivi was wiser than his contempories (and wiser than many bioengineers today) for he realised that practice was more important than theory. The ideas of mechanics provided interesting mental exercise but were seldom of much use to the practising physician. 'We must not be surpriz'd', he wrote, 'to find that the true and genuine Cause of Diseases, can never be found by Theoretick Philosophical Principles.'

Models of man continued to be made. In the age of steam, physiologists made a serviceable model of the body as a heat engine, taking in fuel and using energy in various forms. With the rise of biochemistry, essentially a twentieth century creation, efforts were made to explain the vital functions of the body in terms of chemical reactions. More recently the chemical model has been broadened with ideas derived from physics and genetics in the new science of molecular biology.

The engineer's view of man, which this book seeks to describe, owes something to each of the models already mentioned but is more than a mere symposium. The present age is the first in which the engineer has commanded the insight and the intellectual equipment to make a detailed model of man. Today the bioengineer also has the resources to contrive effective action against many of the defects in design or performance identified by his studies.

The bioengineer uses the ideas and techniques of engineering in two ways. The first is to describe and, where possible, to explain the structure and activity of the body in normal health – an assignment which leads the enquirer into many branches of engineering.

In brief, the engineer sees the body as a self-propelled power source with many ingenious design features. Continuously variable gearing allows a full power output to be developed over a wide range of speeds. The machine is content with any of a great variety of fuels and carries considerable reserve supplies. It embodies elaborate self-repairing facilities and is lubricated for life. The mechanism is controlled by a computer more sophisticated than any electronic device so far developed. Elaborate systems for orientation and navigation sample the environment with devices including a 3-D camera and a hi-fi stereophonic sound reproducer. The propulsion system, using some of the advantages of wheels and tracks, allows the human animal to climb trees, jump ditches and perform many other manoeuvres beyond the capability of any mechanical contrivance.

The brain, originally the control centre for the machine, has developed to allow abstract thought, artistic creativity and, in general, regular or continuous mental activity of a sophisticated kind. This ability is, on the evolutionary time scale, a very recent innovation separating man from the rest of creation.

In studying the mortal machine, we should begin by examining the structure, continuing with a study of the various mechanisms and concluding with a review of the control centre and the possibilities for improvement in design or performance. Among the structural materials, bone is of great interest to the engineer for its successful compromise between strength and weight, its built-in repair facilities and (in the joints) its ingenious lubrication system. The muscles are, in engineering terms, linear motors driven by fuel cells, thus embodying two of the most recent innovations of electrical and chemical engineering. In combination with the bones and joints of the skeleton, the muscles provide a set of low speed reciprocating engines to power the body's physical activities. The other main structural element is the skin. This tough material – the original self-sealing tank – is more versatile than any engineering material. Apart from its protective role in shielding the delicate internal organs and tissues from a hostile environment, it monitors the environment by touch and temperature and, by controlled evaporation, helps to stabilise the internal temperature of the body.

The sense organs, responsible for vision, hearing, taste and smell, have probably deteriorated in modern man, since their survival value is not as great as in primitive times. Their performance is still impressive. The eye has many of the characteristics of a sophisticated camera, but in a much smaller space and with wholly automatic controls. The ear is possibly too sensitive, as evidenced by the prevailing indignation over noise. The ear also has an unnecessarily acute sensitivity to changes in pitch. However, these apparently superfluous capabilities allow great refinement in the perception and appreciation of music — so much so that a room full of electronic equipment is needed to exploit the full potentialities of the combination of ear and brain.

The heart, judged as a pump, excites the admiration of the hydraulic engineer by its subtle design, effortless adaptation to varying loads and extreme reliability. Significantly, the heart-lung machine, which will only take over the functions of the natural mechanism for a few hours, is about a thousand times larger than life.

The kidney is a chemical engineering system, designed and operated with a panache that no engineer would be bold enough to imitate; the artificial kidney is large and clumsy, because synthetic materials cannot approach the performance of the natural organ.

The circulation is an example of a multipurpose system, often found in the body but generally beyond the capability of the engineering designer. The blood is not only a fluid assembly line, delivery network and waste-disposal system but is also the basis of the body's highly effective central heating (and cooling) arrangements and the defence mechanism against infection.

Chemical engineering predominates again in the liver and

the gut, where complex processes are carried out at a low temperature and without any of the powerful reagents which would be necessary in the laboratory.

The brain, which controls and monitors every activity of the body, is a computer with capabilities surpassing those of any man-made electronic system. It is furthermore a computer with the ability to think, to talk, and to engage in artistic creativity.

The engineer is however not utterly speechless in admiration of the human machine. He sees a number of shortcomings in design and performance — some attributable to the change from four feet to two and others which are not so easily explained. With a proper understanding of the natural structure and function of the body, the engineer can help to design spare parts or other devices to support the quality of life when it is impaired by accident or disease. This is the second major responsibility of the bioengineer and is particularly important at a time when technology is not keeping up with the progress of medicine. Unless more of the world's ablest engineers and scientists are brought into the clinical battle, increasing numbers of people will be condemned to mere medicated survival.

2: grit and glue

A sapient schoolboy

The schoolboy who defined a skeleton as 'a man with his inside out and his outside off' was more perceptive than he knew. Many primitive creatures consist essentially of two layers of cells – an inside and an outside. The next stage in complexity is achieved by adding a middle layer from which, during the course of evolution, many organs and tissues have developed, notably the skeleton. The smallest creatures do not need bones. They live, for example, in the sea, which is really a nutritious soup providing nourishment around the clock and taking away waste products without fuss.

As an abode of life, the sea has the great advantage that gravity is almost completely neutralised, so that only the simplest arrangements are needed for moving from one place to another. This advantage is not confined to very small creatures; some kinds of jellyfish are as large as a man and even these can manage perfectly well without any bones.

In general however most animals (leaving aside the tiniest creatures but including birds and fishes) need a skeleton. In man, the most highly specialised of all animals, the skeleton and the muscles attached to it are important tissues for two reasons – structural and functional.

As a structural material, bone is important in two ways. Firstly, the bony framework protects important structures such as the brain, the heart and the eyes. The internal skeleton is of course not the only way of achieving this purpose. Shell-fish have a bony covering which is external to the rest of the body. The disadvantage of this solution is that it does not allow any satisfactory way for growth. The crab is very vulnerable in the interval between shedding one shell and growing another of larger size.

Human hardware

A skeleton is useful also as internal scaffolding, keeping the various organs and systems of the body in the correct relative positions. Most organs and tissues are soft and easily displaced. In an animal which moves as much as man; a chaotic situation would soon appear if the various components were not kept in their proper places. Bones (including teeth) are the only hard tissues which man possesses and they are important in maintaining the shape of the body.

The main job which the skeleton does is concerned with movement. The energy required for movement is of course always derived from muscular exertion. In fishes and reptiles the processes involved are rather simple. But man has the capacity for an enormous variety of subtle and complicated movements, made possible by engineering design of a high order.

A bony skeleton is necessary in the first place to provide the means of exerting effort against the ground, for example in walking. The bones are used also in transforming the effort of the muscle, which is normally a simple contraction, into the elaborate movements that we make at work or at play. As we shall see shortly, muscle anticipates a number of modern achievements in technology, including the linear motor and the fuel cell. The combination of bone and muscle, with the highly developed control system represented by the brain and nervous system, contains much of interest to the engineer.

Lightness and strength

We shall, for a start, consider the design specification for a skeleton. The bones, which are the basic elements in the skeleton, must be rigid but not too rigid, since it is better for them to yield to a certain extent under stress than to break. The forces likely to act on a bone, or on any structural material, are of two main kinds, compression and tension. Carry a heavy suitcase and the bones of the arm are put under tension. Jump down stairs and the bones of the leg are put under compression.

Some materials, such as stone, are strong in compression but weak in tension. Wood, on the other hand, is strong in tension but weak in compression, particularly when the force is applied in the direction of the grain. It is of course possible to find materials such as steel which are quite strong both in compression and in tension — but the specification for bone is more exacting, because it has to be very light; the two hundred-odd bones in the skeleton have a total weight of only about 20 lbs. The combination of lightness and strength is achieved in the skeleton by methods which engineers are only now beginning to appreciate and to imitate.

Another important requirement for bone is that it must maintain its function while growing. When a pair of shoes become too small we throw them away and buy another pair. The bones of the foot are, however, growing steadily for eighteen or twenty years and doing their normal work throughout this time.

Thirdly a bone must be self-healing. If a shaft in an engine breaks, we replace it and fit a new one. But if a bone breaks we expect it to mend itself and, after a while, to be as good as new.

Finally, the bones must be able to articulate, that is to fit freely one against another so that force and power may be transmitted. The joints where bones come together must be self-lubricating and capable of operating for many years — preferably about seventy — without attention. (The lifetime lubrication concept has only been fulfilled in practice, by motor car designers, but recently. The springs and axles of many cars still need attention every 5,000 miles, but some have sealed lubricating systems which last for the lifetime of the vehicle.)

Hidden gems

The material which answers to this difficult specification is made in a way which no engineer, chemist or materials scientist would ever have invented. All of the desirable properties specified in the previous pages are achieved with the concoction of grit and glue generally described as bone. The grit, more often called bone mineral, which accounts for about 70% of the weight of bone (or half of its volume), is an inorganic material composed of calcium, phosphorus, oxygen and hydrogen in proportions corresponding roughly to the formula $3Ca_3(PO_4)_2 . Ca(OH)_2$. This material is hydroxyapatite; it is not a true chemical compound. Apatite minerals are well known to geologists; some are used in making fertilisers but others are quite attractive as gemstones. The bone mineral contains a few other substances, including salts of magnesium, calcium and certain minor elements.

The rest of the bone consists almost entirely of collagen, which is, in the main, a mixture of amino acids and constitutes the glue which is obtained when bones are boiled. Collagen is found in all the tissues of the body but bone collagen is unusual because it can be mineralised. A bone, reduced to its simplest terms, is made of collagen fibres to which tiny crystals of hydroxyapatite are firmly fixed. Sometimes the collagen fibres are tangled together but more often a regular pattern can be observed. The fibres are usually bound together, side by side (rather like a microscopic version of a child's liquorice strap) to form layers or lamellae. In the smooth surface which forms the outside of the normal bone, the lamellae are simply laid one on top of another.

A solid sponge

The interior of a bone is usually composed of osteones, tiny tubes about 15 mm long and ¼ mm in diameter. The osteones are hollow and the interior space is occupied by blood capillaries. The wall of the osteone is formed by lamellae wound in a spiral pattern rather like the stripes on a barber's pole. The wall of a bone is usually a compact material containing a few cavities for living cells, (the osteocytes), connected by tiny channels called canaliculi. When a large bone is cut lengthwise with a saw, some regions show, even to the naked eye, a structure rather like a honeycomb or sponge. The bony structures corresponding to the solid parts of the sponge are known as trabeculae.

Collagen, when separated from the other constituents of bone, is rather like a stiff jelly. Bone mineral is a brittle crystalline material. How does this unlikely combination give bone its unique mechanical properties? In mechanical strength, bone is vastly better than brick or concrete, appreciably better than wood and many metals and not much inferior to steel; indeed on a weight for weight basis bone is considerably stronger than steel.

These properties cannot be explained by regarding bone merely as a kind of rock – for stone is weak in tension, whereas bone has considerable resistance to stretching or bending. Another proposed explanation is that bone is a pre-stressed material in which the collagen fibres (like the steel rods in pre-stressed concrete) are initially stretched while the surrounding mineral material is compressed. This theory is however altogether too ingenious and does not stand up to mathematical investigation; in any case it is very difficult to see how the collagen fibres could be pre-stressed in living and growing bone.

A more plausible model of bone regards it as a composite material. The idea of using one material to improve the properties of another for structural purposes is of course quite old. The carriage wheels of earlier times were built of wood (which is easily worked) but an iron rim was added for greater strength. Earlier still, the architect who designed the gateway of the Acropolis in Athens (in 437 BC) found that the marble slabs commonly used (for example in the Parthenon, on the same site) were not strong enough to span a width of more than about 8 ft. His design called for ceremonial entrances of up to 20 ft in width. Marble, like other varieties of stone, is quite strong in compression but weak in tension; consequently a long beam (which has to withstand considerable tension) is not practicable. The architect solved the problem (as was discovered very much later) by putting iron rods into grooves in the marble blocks, using cement for concealment.

Fibreglass or tyre rubber

In the technical sense, however, compound materials are those in which the two constituents are more closely associated, in a way which is seldom seen by the naked eye or revealed in any simple tests. Bricks made with straw, builder's plaster reinforced with animal hair and papier mâché are examples of somewhat more sophisticated composite materials, but the first significant modern invention in this field was bakelite.

Bakelite was originally a synthetic resin, quite useful as a glue or as an insulating material but of little use in bulk because of its poor mechanical strength. The inventor, Dr Leo Baekeland, found that the mechanical strength of the material was greatly increased if finely divided fibrous material, such as wood flour, was added to the resin before it had completely set. This mixture could be moulded under heat to give a material which was tough, light and cheap.

Fibreglass, which dates from about 1945, is another familiar composite material. Thin fibres of glass, less than a thousandth of an inch in diameter, are very much stronger than the same material in bulk, and indeed very much stronger than steel. Fibreglass contains a great number of these thin fibres embedded in a synthetic resin. The resin, though tough, has rather little mechanical strength. Glass fibres, though exceptionally strong in tension, are naturally easily bent and therefore not suitable for structural purposes. The combination of glass and resin gives a material which is very much stronger than a lump of glass or a lump of resin of the same dimensions.

Glass is a brittle material because, when a crack starts (for example when the surface is damaged) it spreads easily and often results in severe damage. The harmful effect of a crack is due largely to the very great concentration of stress near its tip. Often the damage does not go very far in a metal, which can stretch so as to relieve the stress near the tip of the crack. In its normal state, glass is almost always covered with fine surface cracks which have a distinct weakening effect. Thin fibres, for reasons which are not absolutely clear, are usually free from surface scratches and therefore much stronger.

If fibreglass is stressed sufficiently, the fibres may certainly fail. A crack will however not spread very far. When it reaches the edge of the fibre in which it started, the stress will be relieved by the stretching of the surrounding resin.

In bone, the collagen acts like the resin and the mineral crystals act like the glass fibres. The analogy is perhaps not very good, since the bone crystals are only about 200 Angstrom units long (two millionths of a centimetre) whereas the fibres in fibreglass are usually a few millimetres long.

Another composite material which might be used as a basis of comparison with bone is the filled rubber used for making motor car tyres. In this material the rubber is mixed with carbon black — in other words, soot. This mixture wears much better than plain rubber and can be ten or twenty times better in mechanical strength. Bone is more like fibreglass than like tyre rubber. In some ways its design is even better than that of fibreglass.

Though broken bones are not uncommon in the modern world, a bone will stand up to very rough treatment and usually breaks as a result of severe sudden impact. We have already seen that even apparently solid bone contains cavities for living cells. Probably these cavities have another function in preventing the spread of cracks. For the stress concentration to remain at a level sufficient to tear the bone apart, the tip of the crack must be very narrow. When a crack reaches a cell cavity it is immediately opened out and the stress concentration falls to a level which the surrounding bone can probably withstand. This mechanism has been verified by laboratory experiments in which a chisel was used to crack pieces of bone which were then sectioned for examination under the microscope. In a significant number of instances the cracks ended in cell cavities.

Tension and compression

Bone is a good material for the job that it has to do, but that is not the whole story. In the body, as in most machines, saving of weight is very desirable, since it saves fuel and improves efficiency.

Weight could be saved by making the bones thinner, but this would not be a good solution since the mechanical strength would be greatly impaired. A better idea would be to have the inside of the long bones hollow. This, it turns out, makes very little difference to the mechanical strength.

If a metal bar, a wooden beam or a bone is loaded at one end, the upper surface will be stretched and will therefore be in tension while the lower surface is in compression. This means that somewhere between the upper and lower surfaces there must be a layer which is neither in tension nor in compression. In a solid beam there is in fact a region above and below the neutral layer which is subjected to relatively small forces of compression or elongation when the beam is loaded. Consequently this part of the beam can be removed with very little effect on the mechanical strength.

Why bones are hollow

Many of the long bones in the body are indeed hollowed out. In this respect, bone demonstrates the design principles familiar in tubular steel furniture. There is a further refinement in that the wall thickness is greater near the centre of the bone than near the ends. Here again the natural design conforms to sound engineering principles, because the forces of tension and compression are greater midway along a beam than in the regions near the ends. In trabecular bone, it is not difficult to pick out the regions which have to stand up to tension or to compression. In a section cut through a thigh bone (or femur) the arrangement of trabeculae corresponds closely to the stress lines which an engineer would deduce from the shape of the structure and the load which it bears.

For a typical long bone, such as the femur, the wall thickness is about half of the radius of the shaft. The bone is therefore about 25% lighter than a solid structure of the same outside size, but has virtually the same mechanical strength. In birds, where weight restriction is even more important, some bones have walls which account for only 10% of the total diameter; a thin tube of this sort is liable to buckle under stress, but is usually strengthened by internal struts.

There is really no empty space inside the human bone. The cavities in long bones are filled with marrow, which is the source of red blood cells. The new cells are easily distributed in the circulation; living bone is very different from the material found in a prepared skeleton or on the beach, for it is richly supplied with blood. In man, no bone cell is more than 0.1 mm from a blood vessel.

Business as usual during alterations

An engineer designing a machine or structure can sometimes allow for the possibility of future expansion by the addition or replacement of certain parts. He would however be baffled by the problem of designing a machine which grows continually for twenty years, without any shut-downs, which never needs any maintenance and which can repair itself when accidentally damaged.

Many tissues in the body are able to grow simply by adding on fresh cells and moving outwards at the edges. This method is quite suitable where no great stresses are involved. In sharks and lampreys, the skeleton is composed of cartilage (more commonly known as gristle) which is tough, flexible and well-suited to a placid life in the sea, where the effects of gravity are almost completely neutralized.

In man and many other animals the embryonic skeleton is formed of cartilage but is fairly quickly converted to bone. In the newborn human infant, the skeleton still contains a good deal of cartilage which is gradually converted – though some of it remains throughout life, for example in the external ear, the tip of the nose and the ends of the ribs, where they are joined to the breast bone.

A bone cannot increase its length merely by adding on fresh material at the ends. In many important bones, the ends have to maintain smooth, precise and freely moveable connections (or articulations) with other bones. The increase in length must therefore take place from within and this is done in an ingenious way. As we have seen, the original templates for the bones of the body are made from cartilage at an early stage in embryonic development. The cartilage is steadily replaced by bone – but not completely. In early life, the end of the bone (including, where appropriate, the smooth articulating surface) is separated from the rest of the bone by a thin layer of cartilage, which survives until the bone is fully grown in adult life. The layer of cartilage grows continually. As fresh cells are added, the older cells are invaded and replaced by new bone produced in the shaft. In this way, the growth effectively takes place not in the end cap (known as the epiphysis) but in the intermediate layer of cartilage. This cartilage, which, unlike bone, can grow in a straightforward way, provides the framework within which new bone can be deposited.

When a bone is fully grown, the epiphysis finally joins with the shaft. This process begins at about the age of fifteen (in one of the large bones of the foot) and is not completed until about the age of twenty-five when the collar bone fuses with the breast bone; for other bones in the body, the union of the epiphyses occurs at intermediate ages, a situation which often allows the forensic expert to make a reasonably accurate estimate of the age of a victim.

A bone must increase in thickness as well as in length. New bone is usually produced near the outside and the wall thickness is adjusted by the disappearance of bone from the inner wall of the central cavities. A more complicated remodelling is needed at other places, particularly in bones which are wider at the ends than at the centre, in order to maintain the correct shape during growth. All of these processes involve the activities of two types of cells — the osteoblasts which are responsible for the deposition of new bone and the osteoclasts which are able to destroy unwanted bone, reducing it to substances which are carried away by the circulating blood.

Bone, as we have seen, has a combination of strength and toughness well adapted to its role in the body. Like a man-made structure of good design, a bone has a margin of safety which is adequate for most practical purposes — but size and weight are also important and the margin does not cover every hazard or emergency.

As good as new

Unlike a man-made machine, bone is equipped with its own versatile repair arrangements, which are always ready for action, give round the clock service and remain on the job until the damaged member is restored to a state at least as good as new.

When a bone is fractured, three sorts of damage are done. Firstly, the bone itself may be cracked, split apart or broken into several pieces. Normally, a bone is kept in position by delicately balanced forces exerted by two or more muscles. After a fracture, this balance is upset and the broken pieces of bone may be moved out of alignment. Healing takes place more successfully if the broken ends are brought together and held there with the help of a splint or plaster cast.

Secondly, bone is penetrated by a great number of blood vessels, some of which are torn in a fracture. Thirdly, if the injury is rather violent, the muscles may be torn.

The first stage in the repair process is the formation of a blood clot, which serves two purposes, firstly to seal off the bleeding vessels and secondly to hold the broken ends together while the repair process is being organised. During the first two weeks after the fracture, the blood clot is replaced by cartilage and new bone begins to appear on the inner and outer surfaces of the bone a little distance from the fracture. This bone spreads over the ball of new cartilage around the fracture site and begins to invade it. The cartilage is gradually broken down and replaced by fresh bone. When this process is completed, after two or three months, the site of the fracture is marked by a mass of new bone which extends right through the shaft and bulges over the broken ends. In the further stage of healing this bony mass (or callus) is reorganised by the removal of surplus material so that the final result conforms reasonably well to the original outline of the bone. The whole of this healing process takes several months, but if all goes well the new bone is as strong as the original or even stronger.

Some bones, such as those in the leg, carry heavy loads in normal life and must be kept out of service during the healing process. Occasionally the situation is even more difficult. The head of the femur (or thigh bone) is a hemispherical dome, offset from the main shaft to which it is connected by a short neck. If the neck of the femur is broken, the two fragments are sometimes separated by the action of the muscles and will not heal naturally. An effective method of dealing with this situation is to expose the bone surgically and to drive a steel nail into the bone so as to join the broken-off portion to the main bone. This nail holds the bone together while healing continues and is left in the body permanently.

A nail is perhaps an unkind description of a device which is considerably more than a mere iron rod. An ordinary nail would do too much damage in its passage through the bone and, having a smooth surface, would not always grip very effectively. The Smith-Petersen pin, commonly used for securing the neck of the femur, is a fluted structure with a flat head, allowing it to be hammered into place, and three ribs which can be driven through the bone rather more easily than a pointed rod.

Sometimes a fracture simply refuses to heal. The bone near the broken ends is partially absorbed, leaving a gap. In this situation the best substitute for bone is, oddly enough, bone. A piece of bone taken from some other part of the body or from a bone bank can be shaped appropriately and fitted closely into the gap. The bone graft − an old established and non-controversial type of transplant − acts as a scaffolding through which new bone grows from either end.

The healing of a broken bone is an extreme example of response to stress. Bone displays a varied and impressive capability in dealing with stress − not surprisingly, since it is the body's only hardware. The stresses which bone has to endure are not usually so spectacular as in accidental fracture, but even in normal daily life the forces acting on bones in the leg can reach levels of three or four times the body weight. When a suitcase or heavy parcel is carried in one hand, the hip bone on the opposite side may have to withstand a force of five or six times the weight of the body. In athletes, the stresses can be very much greater, particularly at the moment of takeoff or landing or in such activities as weight-lifting.

An engineer, warned that the normal design requirements might be greatly exceeded, would take the precaution of strengthening critical parts of the structure and this is what happens in the body.

How to see through a bone

Other things being equal, the strength of a bone depends on its mineral content, that is on the amount of hard inorganic material which it contains. The mineral content of a bone may be estimated by reliable methods developed in recent years; gamma radiation from an isotopic source passes through the bone and is detected by a scintillation counter. The extent to which the radiation is absorbed gives an indication of the bone mineral content.

Tests of this kind, made on the thigh bone (or femur), show that bone mineral levels are distinctly higher in athletes than in normal people of the same age. Weight-lifters and discus-throwers showed the highest bone density, followed by runners and footballers; in all of these groups, the most active members showed the greatest increase in bone mineral content. As might be expected, swimmers showed no noticeable difference in bone density.

The strengthening of bone which occurs in response to increased stress has its counterpart in the wasting process observed when normal stresses are removed.

Lost in space

A patient lying in bed for a few weeks will lost quite an appreciable amount of calcium from his bones, though the deficiency is quickly restored on returning to normal activity. Astronauts, though not exactly idle, are also in danger of weakening of the bones through loss of calcium and other mineral contents. During the Gemini IV mission in June 1965, James McDivitt and Edward White lost more than 10% of the mass of some of the bones in their hands and feet. In the Gemini V flight which lasted for eight days, losses reached more than 20% in a few bones.

During these two missions, the crews were not very conscientious in eating the scientifically prepared but unappetising meals prescribed in the flight plans. Consequently their daily calcium intake was lower than normal. The crew of Gemini VII, whose mission lasted 14 days, were subjected to somewhat sterner discipline at meal times and were also obliged to carry out a programme of exercises by pulling on a handle attached to an elastic cord. These measures were quite successful, because the bone losses of the returning astronauts were much smaller than in the earlier Gemini flights.

The effect of exercise on the composition of the bones was confirmed by further experiments in which healthy students spent two weeks resting in bed and enjoying a normal diet. They lost significant amounts of bone mineral. After a few months of normal life, they went to bed again – but this time with a regular programme of exercises similar to those prescribed for the astronauts; the bone losses on this occasion were considerably reduced.

All of these findings, on athletes as well as astronauts, illustrate the principle first asserted by Julius Wolff as long ago as 1892. A modern statement of Wolff's Law is:

The form of the bone being given, the bone elements place or displace themselves in the direction of the functional pressure and increase or decrease their mass to reflect the amount of functional pressure.

The first few words in this definition are important, since it is known (from tissue culture experiments) that bones will develop their characteristic shape and structure even in complete isolation from the rest of the body.

Wolff's Law does not tell us anything about the mechanism by which a bone responds to stress, but recent work on this problem has led to some intriguing speculations and conclusions.

Bone demonstrates the property of piezo-electricity, by which electric currents are produced in response to mechanical pressure. This effect is widely known, for example in the crystal pick-ups used in record players; as the stylus follows the fluctuations impressed (in the recording studio) on the spiral groove of the disc, the pick-up, subjected to corresponding compression and tension, generates electric currents, which after amplification, emerge from the loud speaker to reproduce the original sound.

If an isolated strip of bone is mounted as a cantilever, fixed at one end and loaded at the other, the upper surface becomes positively charged and the lower surface negatively charged. The reverse effect, in which the passage of electric current produces mechanical stress, has also been demonstrated in the laboratory. In further experiments, using cats as experimental subjects, the piezo-electric effect has been demonstrated in the bones of the leg during normal walking.

The role of piezo-electricity in the life of bone has recently been reviewed by Dr Andrew Bassett of Columbia University, New York. He refers particularly to the osteocytes, the living cells which are distributed throughout a bone but whose purpose is not fully understood. It is known that if the osteocytes die the surrounding bone is likely to be absorbed and, in due course, replaced by fresh bone. The nutrition of the osteocytes is therefore important for the health of the bone. These cells are however almost imprisoned by the hard mineral substance in which they are embedded and it is not certain that the tiny channels providing their only communication with the outside world are enough to ensure an adequate supply of oxygen and other necessary materials by the normal process of diffusion.

Bassett suggests that the bone, if it is to remain healthy, must be continually under stress. The resulting piezo-electric currents stimulate the cell membranes to a pumping action which greatly increases the flow of nutrient fluids through the narrow channels in the bone.

Night starvation

If the normal stresses are removed, as when a patient takes to his bed or when an astronaut spends several days in a weightless environment, the piezo-electric activity is greatly reduced and the nutrition of the bones is consequently impaired.

When however the bone is subjected to more than ordinary stress, the supply of nutrient materials (from the circulating blood) is augmented and further growth of bone material may occur in order to produce a stronger structure. These studies and speculations on the piezo-electric effect lend support to the claims (not so far generally accepted) that the passage of an electric current speeds up the healing of bone after a fracture.

In seeking explanations for the remarkable mechanisms which regulate the growth and strength of bones, we have so far given a good deal of attention to athletes, astronauts and invalids. It is however not necessary to concentrate exclusively on these rather specialized groups, since we all spend eight hours or so in bed every night. Recent experiments show that the skeleton does indeed lose some of its calcium during sleep. The advertiser's warning about night starvation may after all have some physiological foundation.

Bone is a remarkable substance, displaying properties which the materials scientist can only just explain and which he certainly cannot imitate. The success of bone as a structural material is further illustrated by the way in which the two hundred or more bones of the body are coupled together to provide great versatility of movement, or comfortable stability when we are at rest. Bones come together in the joints, of which five kinds are found in the body.

The irregularly shaped bones which make up the skull are hinged together for a relatively short time in early life. The main purpose of the joints in the skull is to allow for considerable distortion as the head passes along the narrow birth canal before emerging into the world for the first time. Before long, the skull bones are firmly locked together to serve their new role as armour plating for the brain.

The other four kinds of joints all provide a greater range of movement. In engineering terms, the simplest joint is the hinge, as found in the fingers. Here the joints allow bending through an angle of approximately 90° but, except in a very few double-jointed people, fingers cannot bend sideways or backwards to any useful extent.

A ball and socket joint is better than a hinge for smooth operation under heavy loads. A good example is the head of the femur, which has approximately the shape of a hemisphere, fitting into a corresponding cavity in the hip bone.

The elbow joint is partly a hinge (allowing the forearm to be stretched straight out or folded back until the fingers touch the shoulder) and partly a ball and socket joint, allowing the forearm to be held with the palm of the hand facing upward, or rotated until the palm is facing downwards.

Finally we have the saddle joint which links successive vertebrae in the spine. These joints allow considerable movement backwards and forwards as well as from side to side; some rotation is also possible.

Lubricated for life

All of the joints in the body are self-lubricating, but the methods used have only recently been studied and are still not very well understood. The engineer knows that any material can be worn out by friction or by the inevitable heating which accompanies it. Consequently he tries to keep metal surfaces apart by lubrication — but not too far apart, since then they would not be able to do their work in the machine. What the designer usually tries to do is to make sure that bearings are always protected by a thin film of oil.

Many machines will run for weeks or months with only a few drops of oil, because the lubricating effect is quite successful even when the oil is only a thousandth of a millimetre thick. The oil film can often be conveniently maintained by the motion of the adjoining parts of the machine; for example, oil can be picked up by a rotating shaft and used to lubricate the bearing in which it spins. This system is known as hydrodynamic lubrication. An alternative technique more suitable for heavy loads and low speeds is hydrostatic lubrication. Here the oil is forced under pressure into the space between the two surfaces, so that an effective film of fluid is maintained even when there is no motion. In both of these processes, the lubricating action depends on the presence of a distinct film of oil between the opposing surfaces – and is largely determined by the physical properties of the oil, such as its viscosity.

A different situation occurs when the bearing has to carry high loads with little or no motion. Here it may not be possible to maintain a film of liquid lubricant and the two surfaces will, for practical purposes, be in contact. In this situation, known as boundary lubrication, a very thin layer of lubricant (possibly only one molecule deep) is closely attached to each of the bearing surfaces. The lubricating action depends on the chemical properties of the lubricant, such as the size and shape of its molecules.

Before attempting to explain how the joints in the body are lubricated, we might mention some of the more salient facts. The first and most striking feature is that lubrication of joints is very good indeed. The coefficient of friction in many human joints is in the range 0.005 to 0.02, representing a performance which could hardly be bettered by the most advanced engineering techniques. Yet in some ways the joint does not seem to be at all well designed. Usually the two bones in a joint do not fit closely together, but move relative to each other by a combination of rolling and sliding. As we shall see shortly, there is an advantage in this apparent shortcoming.

A watery oil

The hinges and the ball and socket structures in the body are known as synovial joints, because they are entirely surrounded by a tough synovial membrane. The ends of the bones forming the joint are covered by a layer of cartilage. These bones do not fit closely together and, where they are not in contact, the space between them is filled by synovial fluid, a watery liquid rather like blood plasma without the protein content and with one or two other substances added. Synovial fluid, when removed from animal joints and tested in the laboratory, gives a pretty poor performance as a lubricant. However this failure is not really surprising, since engineers do their tests with bearings made of metals, plastics or other hard substances most unlike the flabby materials from which our bodies are made.

Knowing the mechanical properties of synovial fluid, the loads endured by the joints and the speed at which the bones move (seldom more than 10 cm/sec) it is not difficult to show that hydrodynamic lubrication is out of the question in biological environments. Calculations show that the film thickness in a typical human joint would be only about a ten millionth of a millimetre. Unfortunately the surface irregularities on a typical piece of cartilage are about a thousand times greater. Consequently hydrodynamic lubrication with synovial fluid, however attractive in engineering terms, would not keep the opposing bones apart and would be a complete failure in practical terms.

Boundary lubrication seems more plausible — but the frictional forces measured in human joints are five or ten times below the lowest limits achieved in any engineering materials with boundary lubrication.

Weeping and gnashing

An attractive theory, which has been developed recently, suggests that joints display weeping lubrication, a technique previously unknown to engineers. Cartilage, when examined with the electron microscope, shows a structure rather like a sponge with very fine pores. Normally the pores are filled with synovial fluid. When one bone in the joint bears on the other, the pressure squeezes synovial fluid out of the cartilage to act as a lubricant. This process resembles the hydrostatic lubrication system mentioned earlier, with the added advantage that the lubricant is released only when it is needed, and in amounts corresponding to the load on the joint.

The question which naturally arises is: how does the cartilage keep itself charged with fluid if it is constantly under pressure? We may now see some purpose in the apparently careless design of joints with two bones not at all well matched to each other. If the bones were a perfect fit, the cartilage would indeed be drained of synovial fluid and the lubricating action would be exhausted. What happens in practice however is that the region of contact between the two bones changes because of the sliding and rolling action as the joint is used; even when we are apparently at rest, the bones are usually moving very slightly, thereby avoiding drying out of the cartilage. As one bone in a joint moves over the other, the squeezed cartilage has a chance to recover and to soak up synovial fluid in readiness for its next load-bearing duty.

The study of joint lubrication, so long neglected, is now a matter of active interest to engineers and surgeons. The problem is of great potential importance, because the rheumatic diseases from which all of us will suffer, if we live long enough, may reasonably be regarded as failures in joint lubrication.

Spare parts

Meanwhile, joints continue to break down because of imperfect lubrication or more serious structural defects. Heroic measures are quite often necessary by way of repair. When faced with a problem of failure in a bearing, an engineer would instinctively decide to replace it. This prescription is, however, very difficult to achieve successfully in the living subject. The surgeon can, without great difficulty, expose the faulty joint and, if necessary, remove it. The trouble is that the engineer seldom knows what to put in its place. One problem is posed by the lack of suitable materials. In general, it is difficult to find any substances which are fully compatible with the tissues and fluids of the body. Apart from this, engineers have not yet found any lubricating system as neat and as effective as that provided naturally by cartilage and synovial fluid.

The biological ball and socket joint provides the most formidable challenge to the engineer, since it is at the junction of the hip and thigh that the largest stresses are encountered. As long ago as 1890, Thomas Gluck, a German surgeon, used artificial joints made of ivory, fixed to the rest of the skeleton by nickel plated steel screws, sometimes reinforced by glue. More recently, replacements of the ball or socket (or both) have been made from steel alloy or from various plastic materials.

The fixing of the replacements to the adjoining healthy bone is always a difficult problem. Rather surprisingly, a reasonably satisfactory lubrication is usually provided by body fluids. If only one element of the joint (that is either the ball or the socket) is replaced, the natural lubricating system is not completely abandoned since there remains one surface of cartilage, able to produce synovial fluid as in a normal joint. Even where both the ball and the socket are made of metal or plastic material, fluids from the surrounding tissues are often able to provide adequate lubrication.

In engineering terms, the replacement of bone by metal is not very attractive. The elasticity of steel is ten times that of bone. This means that, under any given load, a piece of bone will stretch ten times as much as a piece of steel. For this reason, it is difficult to achieve a close and permanent union between bone and steel, particularly in the femur where such heavy loads are endured. In recent years, materials scientists have been able to imitate nature more closely by developing plastic materials, reinforced with glass or other fibres to give a product fully as strong as natural bone and with nearly the same elasticity. It is of course still difficult to find any synthetic material which can be left in the body for many years without deterioration. It is reasonable to expect that these difficulties will respond to further research and that a good replacement for bone will eventually be found.

3: six hundred engines

Motion without wheels

Many of the systems of the body can be described without much difficulty in engineering terms. The structural engineer recognises familiar problems in bones used as girders, bridges and pillars. The hydraulics expert appreciates the ingenious ways in which the body's fluids are circulated and conserved, while the heating and ventilating consultant admires the efficiency and economy of the nose. The control engineer sees many feedback loops and other examples of automation and the computer designer can use his experience in trying to understand the brain.

For the mechanical engineer, however, the study of man presents some difficult problems. The body is, among other things, a workshop, chemical laboratory and data processing plant. It provides its own power for all of these purposes and is, in addition, freely mobile with a built-in propulsion unit.

Most engines, whether used for propulsion or for industrial purposes, fall into one of two categories, reciprocating and rotating. The early steam engine, with its piston moving in a cylinder, was a simple and effective reciprocating engine, generally used for pumping water – an activity in which the up and down movement of the piston proved quite convenient. With the addition of mechanical devices such as the crank and the fly-wheel, a to and fro movement can be converted to rotation, which is more useful for driving machinery in a factory or for transport; the modern internal combustion engine, as used in the motor car, starts as reciprocating engine but converts its energy to rotary movement.

The engineer looking at the body's power system is impressed by two features. Firstly he sees no rotary movement anywhere. The wheel, so convenient in almost every form of transport and every kind of machine, simply does not occur in nature.

There is, of course, a reason for this apparent shortcoming. The engine of a car, a boat or an aircraft is quite separate from the rest of the structure (it can, for example, be removed with little difficulty) and can generate rotary movement without inconvenience. But the design of the body requires every part to be connected to its neighbours by blood vessels, nerves and other tissues. Consequently, rotation is possible only up to one complete turn and must then be reversed. Lacking the possibility of rotation, the body uses reciprocating motion for transport and for every other form of physical activity.

In many ways this is a better solution than a wheel. It is true that a man who walks at 4 mph could travel twice as fast by applying the same muscular effort to drive the wheels of a bicycle. But the wheel requires a hard and reasonably level surface; for the steep or rough ground which covers much of the globe the feet are better. The ability to adjust the length and inclination of the leg to the terrain obviates the need for elaborate shock absorbing devices which a wheeled vehicle needs even on the highway. The second surprising design feature which the engineer notices is that, instead of a single engine, the body has more than six hundred. These are the muscles, each adapted to its particular purpose. Some (in the leg for example) are concerned with movement and others (particularly in the arms and trunk) are involved in different forms of work. Tiny muscles focus the eye lens and others adjust the sensitivity of the ear for protection against loud noises. Muscles are involved in eating, drinking, digestion and breathing; the heart is a highly specialised muscle. Muscles make up about 40% of the body weight in man, and a little less in woman. The flesh that we eat from animals, birds, and fishes is nearly always muscle.

As a material for building an engine, muscle does not look very attractive. True, the heart muscle can do its job by squeezing the blood out into the arteries and relaxing to let it flow in again — but a jellyfish cannot stand up and it is not obvious that muscle alone is capable of doing a wide variety of useful work. The secret is in the association of the muscles with the bones. In effect, the body is a complex meccano-like structure with a large number of stiff rods, loosely jointed together. If the nuts and bolts forming each joint are tightened, a rigid immobile structure is produced. On the other hand if the joints are left loose, the whole assembly falls in a heap on the ground. What the muscles do is to selectively tighten and loosen the connections so that the structure changes its shape and position according to the job that has to be done. A skeleton cannot be balanced on its two feet, and a muscle once removed from the body is no more than a flabby lump of meat — but the combination in a living person presents a machine of remarkable power and delicacy.

If the outward appearance of a muscle does not seem well adapted to its purpose as an engine, detailed examination is even less encouraging. A typical muscle is a rope made of jelly, a multitude of thin fibres. In carrying out their work, muscles use only one simple mechanical action, the contraction of fibres. This means that a muscle can pull (or squeeze) but can never push. In practice however the combination of muscle and bone allows a very wide range of movements and forces.

There are two main kinds of muscle, smooth and striped. Most of the muscles that we can see or feel are of the striped variety; the stripes are only visible under the microscope. Striped muscles are found in the arms, legs and trunk and are all under voluntary control; in walking, eating or reading a book we command the appropriate muscles by conscious effort of the will.

There are however a great many other muscles over which we have no conscious control. These are the smooth muscles found in the stomach and in the intestines (where they churn and propel the food, in the walls of veins and arteries (where they regulate the blood pressure and flow), in the lungs and in the wall of the bladder.

These muscles are permanently under automatic control. We do not, for example, have to make any conscious effort to keep breathing — otherwise we should not be able to sleep. It is of course possible by conscious effort to hold the breath but nobody ever commits suicide in this way, because the increasing concentration of carbon dioxide in the lungs triggers an automatic mechanism which overrides the conscious desire to stop breathing and forces the muscles of the ribs, diaphragm and lungs to move so that fresh air is taken in. Children can sometimes hold their breath long enough to pass into unconsciousness, but the same automatic mechanism brings them round again. Smooth muscle is designed for slow steady effort and, in normal health, never becomes fatigued. Striped muscle, on the other hand, is capable of a much wider range of exertion and lets us know when it has had enough.

Fuel efficiency

In reviewing the action of muscles we have to consider two aspects, the chemical and the mechanical. The chemical aspects of muscle are concerned with the fuel that it uses. A muscle is an engine, but it is not a heat engine in the sense familiar to engineers. Steam engines or internal combustion engines burn fuel to produce heat. This heat is then converted into work — not completely, but usually at a rather low efficiency, so that the final result of burning the fuel is a small amount of work and a larger amount of heat.

A muscle operates in a more sophisticated way, by oxidising a specialised fuel which is manufactured in the body. The chemical energy released in this way is converted partly into work and partly into heat. In this respect the muscle is not much like the steam engine but resembles the fuel cell, a recent invention in which fuel (for example, hydrogen) reacts with oxygen to produce electrical energy directly, with heat as a by-product rather than (as in most engines) an essential intermediate stage in the production of useful energy.

The efficiency (that is, the proportion of the chemical energy of the fuel that is actually converted to work) is a useful way of comparing one engine with another. The calculation is rather difficult for a muscle, but gives a rating of about 25%. This is better than a steam engine, about the same as a car engine and not as good as a diesel engine. Muscle is, however, better than all of these in one important respect. It works without reaching a high temperature. No heat engine, however well designed, could reach an efficiency of 25% unless it ran at a temperature of at least 100°C.

Another useful index of engine performance is the power to weight ratio. A man can work at the rate of 6-7 horse power for a second or two (for example in weight-lifting) and an athlete can maintain about 2 hp in a 100 yard race. For prolonged effort, up to an hour or so, the limit is about 0-5 hp. A well-designed engine of the same weight as a man could produce about 50 hp.

The comparison is not quite fair because much of the weight of the body should be regarded as payload and not included in the weight of the engine. Remembering that the human engine starts instantly, adapts automatically to changing loads and is not very fussy about fuel, the performance is impressive.

Most engines, including fuel cells, have to be fed with the proper fuel. Muscle is no exception, since it will only work when supplied with one particular fuel, a chemical known as adenosine tri-phosphate or ATP. The body's digestive system and chemical processing plant can make ATP from the constituents of any normal diet.

A car can carry enough petrol for several hours' driving but the body manufactures fuel for the muscles as the need arises. The amount of ATP normally present in the muscles would only support a few seconds of exertion. If the amount of

ATP falls, the muscle becomes stiff, as for example in rigor mortis. There are however three ways in which the supply of fuel can be maintained: (a) When ATP gives up its energy for muscular contraction, it is converted to ADP (adenosine diphosphate). The ATP is regenerated by the combination of ADP with phosphocreatine, another compound always present in the muscles. (b) Energy needed to restore the reserves of phosphocreatine comes from the oxidation of glycogen; this is a carbohydrate and a low grade fuel present in quite large amounts in muscle. Its utilisation needs oxygen. (c) When the supply of oxygen is not sufficient, as can easily happen during vigorous exercise, glycogen undergoes an alternative process, being converted to lactic acid and releasing a moderate amount of energy.

In practical terms, the significance of these reactions is that a useful amount of energy is always available for a burst of activity. This energy may be used quickly in sprinting for a train or more slowly in a quarter-mile trot. When it is gone the only other source available is the oxidation of glycogen which can only proceed at a rate dictated by the availability of oxygen carried by the blood to the muscles. In a trained athlete a steady power output of 0.4 or 0.5 hp can be kept up for a long time. The sprinter, using the short-term energy reserve, can cover a hundred yards at a speed of 23 mph; the world's record for a two-mile run corresponds to a speed of only 14 mph but the ten mile record represents a speed of almost 13 mph.

In many animals there is a visible distinction between muscles adapted for prolonged steady exertion (requiring a good supply of oxygen) and those designed for short bursts of activity, using material such as ATP and phosphocreatine which are already present in the muscle and can release energy without needing oxygen. Only a small proportion of the oxygen in a muscle is combined with haemoglobin, the material which gives red blood cells their colour; most of it is combined with myoglobin (a substance rather similar to haemoglobin) and cytochrome, another coloured material.

The breast and wings of a chicken provide white meat, because the farmyard fowl (or broiler bird) seldom flies for more than a few yards. A game bird, such as a pheasant or partridge, has to fly more seriously; its breast and wing muscles need a good oxygen supply and are therefore darker in colour. Similarly the rabbit, which runs back to its burrow but seldom further, contains mostly white meat while the hare, which lives in open country and has to cover much greater distances, provides dark meat.

The energy overdraft

When an engineer designs a machine he calculates the maximum amount of work that it has to do, adds a percentage for safety and provides a power supply accordingly. For the human body, a calculation of this sort is not of much help. If the muscles and oxygen supply network are designed for eating, sleeping and sitting at a desk, they will not be able to provide the extra effort needed for running upstairs. On the other hand it would be wasteful to organise the body's muscular engines for maximum effort all the time.

The solution adopted is very ingenious and is better than anything that a mechanical engineer can do. In brief, the muscles are allowed to use oxygen faster than the lungs and circulation can deliver it – in other words to incur an overdraft. When the athlete subsides panting at the end of a race, breathing deeper and faster than he would normally do when sitting at rest, he is repaying the oxygen debt.

This arrangement is important because the lungs are not very good at capturing oxygen from the air; to put one litre of oxygen into the blood we need to take in about twenty litres of air. An athlete would need seven litres of oxygen to provide the muscular effort for a hundred-yard sprint. During the ten seconds or so needed to run this distance his lungs cannot deliver much more than about half a litre of oxygen to the blood. A sprinter would run as fast if he held his breath during the ten-second dash – as some athletes in fact do. The muscles do not give up when they have used all the available oxygen; they continue to work and, in the process, accumulate lactic acid and other substances which can be tolerated for a short time. At the end of the race the oxygen load is repaid.

In the early stages of a long race, the accumulation of lactic acid in the muscles can be quite painful and the effort to take in enough oxygen can test the athlete's endurance severely. After a while, however, he feels better adapted to the task, having gained his second wind. There is no completely satisfactory explanation for this change but it is probable that increases in heart rate, blood pressure and respiration rate help to deliver more oxygen to the muscles, so that the supply reaches an equilibrium with the demand. A long distance runner will, however, still have to repay the oxygen debt at the end of the race because the body is incapable of taking in oxygen surplus to its immediate needs.

The biological engine

A muscle, as we have seen, is an engine — a device for doing work. The five essential features of an engine are:

(a)   the fuel supply
(b)   the exhaust
(c)   the cooling system
(d)   the mounting
(e)   the control system.

Each of these features can be recognised in a muscle. Every muscle is linked with the circulation by an artery and a vein. The artery brings oxygen and other essential chemicals, corresponding to the fuel intake of an engine. The vein takes away waste products, serving the purpose of the exhaust pipe. No engine converts all of its fuel into useful work; some of the chemical energy, usually a large proportion, is converted to heat. Most engines need water, oil, air or some other fluid to keep them cool. A muscle generates heat when it is working; that is why exercise makes us warm. The circulating blood is the cooling system, removing surplus heat and distributing it to cooler parts of the body or to the blood vessels near the skin where it is lost by various processes discussed in chapter 8.

An engine must have a firm base if it is to do its work in an orderly way without thrashing about. A stationary engine is usually bolted to a heavy base-plate, fixed to the ground, and a mobile engine is attached to the chassis or body of the vehicle that it drives. In the body, many muscles are securely attached to bones. The connecting link for this purpose is a sinew or tendon, a cord made of tough fibrous tissue; the tendons behind the knee can be felt quite easily when the thigh muscles are flexed.

The controls of an engine usually include devices for starting, stopping and regulating the power output. The heart muscle has a built-in control system and beats regularly throughout the whole of life without any conscious effort. On the other hand, the muscles attached to the skeleton and used for most of the activities of the body do nothing at all unless the brain tells them to contract. The control system, regulating the timing and force of the contraction is all represented by slender nerve fibres. The design and operation of this control system presents some interesting problems.

The chain of events involved in the contraction of a muscle has been studied intensively in recent years and is now reasonably well understood. Contraction is triggered by the release of calcium, normally kept out of the way in a storage site within the muscle. The increase in the concentration of calcium ions needed to start the contraction can only be brought about by diffusion and this process, as we shall see later (page 113) is effective only over very short distances. Many activities of the body require a very quick response of a muscle to the signal telling it to contract. If this response time is to be kept within the acceptable limit of a few milliseconds, diffusion will be effective only over a range of a few thousandths of a millimetre — not enough to span the

the distance between the surface and the centre of a nerve fibre and still less to carry a message from the brain.

The first part of the design problem is to transmit a signal from the brain to a muscle — a distance which may be a metre or more — in a few milliseconds. The obvious solution is to use an electric current. It would however be difficult to send a current of the normal kind (consisting of a flow of electrons) down a nerve fibre which, like most of the tissues of the body, is not a good conductor. The signal is actually transmitted by the movement of ions in a rather complicated but quite effective way at a speed of up to a hundred metres per second, not as fast as an electric current (which travels at nearly the speed of light) but quite fast enough for the task in hand.

The next difficulty is that the amount of electrical energy which can be passed along a nerve fibre is not nearly enough to stimulate a muscle fibre, which is very much thicker. This is a problem familiar to telegraph and telephone engineers, who deal with it by inserting an amplifier in the line. At the junction of a nerve fibre and a muscle there is a chemical amplifier. Arrival of the feeble electrical impulse down the nerve causes the release of acetylcholine, a chemical which effectively stimulates the movement of calcium ions and the subsequent contraction of the muscle fibre. To make the nerve-muscle junction ready to receive further impulses, the acetylcholine must be removed as soon as the contraction has been effected. This process is usually accomplished by the action of cholinesterase, an enzyme always available at the junction. Some of the nerve gases made for biological warfare are cholinesterase inhibitors, that is, they prevent the cholinesterase from doing its job and therefore block the transmission of nervous impulses completely. In these cir-cumstances the victim is of course completely paralysed.

Some engines work flat out all the time, but usually there is a means of regulating the output. A car engine, for example, is provided with a throttle which, by adjusting the flow of fuel, acts as an accelerator. A muscle is rather like a car engine with a great many cylinders but no throttle. When a muscle fibre receives an electrical impulse, it may do nothing (if the impulse is too small) or it may contract, but there is no intermediate stage. The complete muscle contains many thousand fibres and regulates its output by adjusting the number of fibres which contract and the rate at which they contract. A car engine regulated by firing some cylinders but not others would make rather jerky progress, but this method of control is quite satisfactory for a muscle.

Since a single muscle fibre can generate only a feeble effort, it would be wasteful to have one nerve fibre to every muscle fibre. What in fact happens is that one nerve fibre serves several muscle fibres, which contract in unison as a single motor unit. The tiny muscles which control the movement and focusing of the eye have small motor units with one nerve fibre to about ten muscle fibres, thereby achieving very precise control. In larger muscles which do not require the same degree of delicacy, the motor unit may contain a hundred or a thousand muscle fibres. When a muscle is being exercised continually, some motor units rest while others are active.

Levers

Many of the body's muscles, such as those in the gut, the bladder and the diaphragm, do their work unobtrusively without conscious effort. The skeletal muscles, attached to bones, do their work in a more obvious way. Hold the arms straight out, palm uppermost, bend it at the elbow and raise the hand towards the face. In this movement, the biceps muscle in the upper arm contracts. One end of this muscle is attached to the top of the humerus, the bone of the upper arm, and the other end is attached to the radius, one of the bones in the forearm; the tendon which joins the muscle to the bone can be felt quite easily as the forearm is raised.

In the movement just described the forearm acts as a lever. The lever is a simple machine, widely familiar in daily life, by which a force or effort applied at one point may be made to do work at another point. One form of lever is demonstrated in the scissors, where the effort is applied by the finger and thumb and the load is represented by the object to be cut. A lever always has a pivot or fulcrum – in this case the intersection of the two blades. The nutcracker is a different kind of lever with the fulcrum at one end, the load close to it and the effort further away. The effect of this construction is to amplify the force which can be applied by the hand so as to crack the nut – a task which could not be done by simple pressure of the fingers.

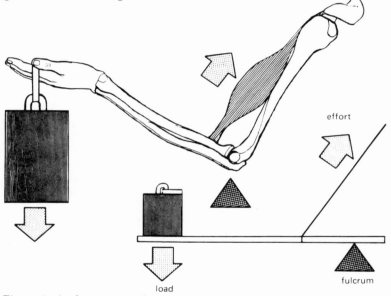

Figure 1 the forearm as a lever

The type of lever most often found in the body is not much used in engineering, for reasons which will soon become apparent. The biceps muscle is attached to the radius at a point quite near the elbow, which acts as the fulcrum of a lever. The effort is therefore applied close to the fulcrum while the load, represented by the hand, is considerably further away. In this situation, raising the forearm is rather like opening a door with its handle close to the hinge – a considerable effort is needed. The force which has to be exerted by the biceps muscle is about seven times the weight that can be lifted in the outstretched hand. The tendon connecting the biceps muscle to the radius is quite strong; although not much thicker than a shoelace it can withstand the tension of a hundred pounds or more. It does however seem strange that the combination of muscle and bone forms a lever system which actually diminishes the effort supplied by the biceps – in engineering terms, a device with a mechanical disadvantage of seven.

There is, however, an important compensation. As the forearm swings about the elbow the hand moves seven times as far (and seven times as fast) as the point of attachment of the biceps to the radius. What is lost in force is therefore gained in speed. Many sports such as boxing, baseball and cricket, and recreations such as playing the piano, depend on the ability to move the hands quickly. A similar lever system in the legs gives man and other animals the ability to run, an ability which had considerable survival value in primitive times and is useful today in city traffic.

In man, the forearm is distinctly shorter than the upper arm, a design feature which represents a compromise between the conflicting requirements of speed and power. In animals the compromise is sometimes different. The adventures of Tarzan are, to an engineer, not altogether convincing. In the ape, the forearm is twice as long as the upper arm, giving a long reach and the ability to move the hand very quickly. A man, with a short forearm, could not swing from branch to branch with the same skill and confidence.

The short forearm has some positive advantages. A few apes can paint, but none will ever produce great art because the long forearm makes delicate control of a brush impossible; clumsiness is the price that has to be paid for a long reach.

Sometimes power is more important than speed, for example in eating. The teeth are attached to a lever system, with a mechanical disadvantage of 3.5 for the incisors (used for light work) and 1.7 for the back teeth, where most of the heavy chewing is done; the back teeth consequently press twice as hard as the front teeth.

An outstanding disadvantage of muscle, in terms of engineering design, is that it can do work only by contracting – in other words a muscle can pull but can never push. We have seen how, by the action of the biceps muscle, the arm can be bent at the elbow and the hand raised to the head. How does the arm return to the horizontal position? If the biceps simply stopped contracting, the forearm would drop suddenly with painful results. In practice, however, the return movement can be carried out very delicately and without any haste or discomfort. The secret is that muscles usually act in pairs, with one opposing the other; the efforts of the antagonists do not exactly cancel out but result in a smooth controlled movement. In the movement of the forearm which we have just considered, the biceps has its counterpart in the triceps, another large muscle which runs down the back of the upper arm and under the elbow to join the ulna, a bone which lies alongside the radius in the forearm. When the hand is lowered, the triceps contracts (helping to straighten the elbow) while the biceps relaxes. The tensions in the two muscles are automatically adjusted so that the movement is made in a controlled way.

It has already been mentioned that the skeleton is a loosely jointed mechanism without much inherent stability. How are we able to move the forearm alone, without chaotic participation by the bones of the upper arm, the shoulder and the chest? The conscious effort of deciding to move one part of the body brings into action an elaborate stabilising mechanism. Before the forearm starts to move, muscles in the shoulder (working, as always, in pairs) combine to hold the upper arm firmly in place. At the same time, muscles in the neck and chest act so as to immobilise the shoulder. Where a really substantial effort is needed, as in lifting a heavy weight or pushing a spade into hard ground, the whole body becomes rigid so as to provide a firm base for the working muscles.

The body's engines, though built of such unpromising materials as jelly and gristle, do their work well and demonstrate some surprisingly modern engineering ideas. For many purposes, including transport, serious attention is now being concentrated on the linear motor, a recent development in which the circular windings and rotary motion of the electric motor are rearranged with flat coils and straight-line or reciprocating motion. A muscle is essentially a linear motor, driven by a low-temperature fuel cell, and is in most respects appreciably ahead of current engineering practice.

4: the original shrinkwrap

During the early days of radiology, around the beginning of the present century, it was noticed that a radiologist would seldom eat a boiled egg. Fried, scrambled or poached eggs were accepted willingly — but at the sight of an egg-cup the radiologist's appetite disappeared.

This curious change in eating habits was not fully understood at the time. The explanation which afterwards emerged involved the response of the skin to the unexpected challenge presented by the arrival of X-rays, a form of radiation not previously found in nature. Before returning to this problem, we shall look at some of the other interesting properties of the skin.

Broadly speaking, the skin serves two purposes, mechanical and biological. In its mechanical role, skin is the original shrinkwrap, providing a closely fitting cover, carefully adjusted as to thickness, elasticity and surface texture for a wide variety of tasks. In its biological role, the skin is the largest organ of the body, housing the sense of touch and playing an important part in the regulation of heat loss and the control of blood pressure.

The main reason for the skin is that most of the tissues of the body are rather fragile and that all of them (even those which are not themselves fluids or semi-fluids) require a very moist environment and must therefore be protected from loss of water by evaporation.

In its function as a buffer between the internal structures of the body and the external environment, the skin displays many useful properties. First, it acts as a waterproof coat, keeping out rain and bath water. It might be thought that the skin is waterproof in one direction but not in the other, since large quantities of water can be lost from the surface of the body in perspiration. As we shall see later, the two processes are quite different.

The environment of primitive man was, in the chemical sense, a good deal cleaner than the modern world, but the skin is able to cope with most of the chemical insults that we inflict on it, whether by atmospheric pollution or by soap, detergents, deodorants, powders and other materials which, though demanded by convention, put a severe strain on the natural mechanisms.

The thick skin of the fingers, palms of the hands and soles of the feet is a good heat insulator. Skin is also an effective electrical insulator, particularly when it is dry. This property was of no significance in early times but is now useful because of the many electrical appliances associated with modern life. The underlying soft tissues of the body conduct electricity quite freely and, were it not for the insulation provided by the skin, the use of electricity in homes and factories would be much more dangerous.

The electrical resistance of the skin, though important for safety from shock, does cause difficulty in the study of the minute electrical signals generated by nerves and muscles and often useful in the investigation or diagnosis of disease.

Finally, the skin gives mechanical protection from the minor accidents of life. In this matter, its elasticity is a great help since it allows the pressure associated with a blow or scrape to be spread over a large area.

The integrity of the skin as a mechanical barrier is very important in avoiding infection – though it has a thriving population of its own germs, none of them really harmful and some quite useful.

Our two skins

Before exploring the properties of skin in more detail, it will be worthwhile to describe the structure of the remarkable material with which we are dealing. Faced with the problem of designing an outer jacket for the body, the engineer would immediately recognise two conflicting requirements. Since the skin has to be manufactured from materials available within the body, its inner surface must be composed of living cells, able to extract oxygen and various structural materials from the circulating blood. The outer surface must however be much more robust, since it is in contact with an external environment not at all favourable for the survival of unprotected living cells.

The solution is ingenious. Our covering is made up of two layers. What we generally regard as the skin is really the outer layer, known technically as the epidermis. This consists of a few layers of living cells of a simple kind, with a top layer (the stratum corneum) composed of dead cells, flattened

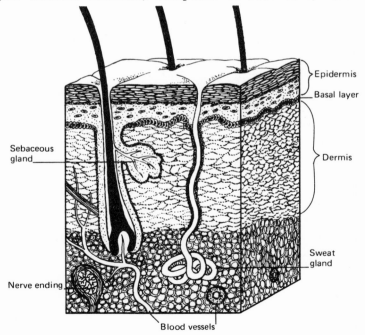

Figure 2 the skin

together to form a tough surface. This layer is not thicker than tissue paper over most of the body but may reach 1 mm or more on the soles of the feet and, in manual workers, on the palms of the hands. The epidermis is essentially a structural membrane without any blood vessels; it does however contain some nerve fibres which are responsible for the sensation of itching. As the cells in the epidermis develop, they move outwards, die and change into the tough surface layer. The dead cells in this layer are constantly being rubbed away by normal wear and tear; several thousand disappear whenever we shake hands or open a door. The effective life of the stratum corneum is about five weeks.

The major part of the skin is contained in the dermis, a more substantial layer which may be 1.5-2 mm thick and is well supplied with small blood vessels, nerve fibres and sweat glands. The upper surface of the dermis is corrugated and the lower (or basal) layer of the epidermis conforms to this pattern. Over most of the body, the outer surface of the epidermis is quite smooth − but on the palms and soles, the ridges persist right through the epidermis, helping to provide a better grip. The fingertips are roughened still further by complicated patterns of whorls and loops. It might be thought that a standard design would be used here to provide enhanced friction − rather like the tread of a car tyre. In fact, for no very obvious reason, the fingerprint pattern is unique to every individual; even identical twins have clearly distinguishable fingerprints.

The dermis is composed mainly of collagen, a versatile protein which occurs also in tendon and in bone. The dermis also contains a good many elastic fibres. Leather is made by chemical treatment of the dermis of animals.

The dermis rests on a layer of subcutaneous fat which improves the insulating properties of the skin and also allows considerable movement of the skin without disturbance to underlying organs and tissues.

The skin is subjected to considerable mechanical stresses; consider for example the stretching and squeezing involved in such activities as drawing a cork or using a screwdriver. The structural integrity of the skin is maintained by a number of ingenious design features. The living cells of the epidermis (in and near the basal layer) are supported by a network of fibres with numerous anchor points formed rather like spot welds, by localized thickening of the walls of adjacent cells. In this way, the stresses produced by the formation of the epidermis, in response to external forces, are distributed over a larger area.

X-rays and fingerprints

We shall now return to the radiologist whom we left at the breakfast table a few pages back. The epidermis, as we have seen, sheds its outer surface continually. This loss is made good by the production of new tissue (by cell division) in the lower layer of the epidermis – a process which, rather surprisingly, occurs at an uneven rate, with maximum growth between midnight and 4 am when most of the activities of the body are at their lowest ebb. The corrugated surface of this layer has the advantage of providing a larger area for cell division and replacement than a flat surface.

In the normal way, cell division proceeds in the epidermis at a rate which maintains the outer layer of dead cells at a satisfactory thickness. If the skin is exposed to X-rays, some of the cells are destroyed. Obviously the outer layer of dead cells cannot be damaged further but the lower layer of the epidermis, where cell division is constantly in progress, is particularly vulnerable.

A small dose of radiation produces no dramatic effect because the rate of cell division adjusts to keep pace. If the radiation exposure is continued, cells in the basal layer of the epidermis are destroyed rather too quickly to maintain the normal equilibrium. The first effect is that the corrugated layer of cells changes into a smooth sheet; cells are, so to speak, joining hands to keep the line intact. Consequently the fingerprints disappear – a process which does not yet seem to have been used as the basis for a detective story. What is worse, the supply of new cells from the basal layer is no longer enough to maintain the epidermis at its normal thickness. Consequently the outside of the skin (now quite smooth) is nearer to the nerve endings in the dermis. These nerve endings are of course quite sensitive to heat. The epidermis of the fingertips is, in a normal person, relatively thick and provides a useful amount of insulation, allowing us to handle rather hot objects without suffering pain. In a person who has been exposed to a considerable amount of X-radiation, the epidermis is so thin that sensitivity to heat is greatly enhanced. That is why the old-time radiologist would never eat a boiled egg; the pain when he grasped it to slice off the top was unbearable.

X-rays are an unusual hazard, but other forms of radiation are very common, and their effect on the skin presents some difficult design problems.

Man, like every other living creature, could not live without radiation. Apart from underground geysers and, more recently, nuclear power, the sun's radiation is the only source of energy that we possess. This energy is often used indirectly in the form of foodstuffs and fossil fuels; even hydro-electric power depends on rainfall which is a consequence of evaporation produced by the sun's heat. The solar radiation is used directly to keep us warm; in the days before clothing, houses and artificial heating, the early civilizations began in regions with an average temperature of about 20°C, which is still the optimum for human comfort.

Quite a lot of the sun's radiation (30-40% for a white man and about 15% for a black man) is reflected from the skin. The rest is absorbed and most of it is immediately converted to heat. Sunlight does however contain a small amount of ultra-violet radiation and most of this is absorbed in the epidermis. Ultra-violet radiation has enough energy to disrupt molecules and to damage living cells. Moderate exposure to the sun provokes a defensive mechanism involving first the migration of melanin (a brown pigment present in the skin) to the epidermis and later the production of further supplies of melanin. This material then shields the skin by trapping some of the incoming ultra-violet radiation before it can cause damage. Negroes have a great deal of melanin in the skin.

The body does not have any very effective defence against prolonged exposure to sunlight, though this hazard has always been present. The long-term effects include thinning of the epidermis and dermis, loss of elasticity, wrinkling and increased hazard of skin cancer.

Waterproofing

Skin of normal thickness is not only a useful insulator but a barrier to almost everything. Were this not so, we should be in serious trouble; life depends on the maintenance of a specialised interior environment which must be preserved from contamination, not only by infectious organisms or other foreign tissues but also by unwanted chemical substances.

The most important single property of the skin is its waterproofing. The circulating blood is, as we have already learnt, a salty fluid rather like the sea water where primitive life began. Many important processes, in plants and animals as well as man, depend on the passage of water or other materials across a membrane separating a dilute solution from a concentrated solution. This process is important in the transfer of oxygen and carbon dioxide between the lungs and the blood. If the skin allowed the free passage of water (as do many other membranes inside the body) we would be in danger of swelling in the bath as water passed through the skin to dilute the blood.

We can, to a certain extent, already cope with this hazard because we drink quite a lot of watery liquids which eventually pass into the circulating blood. It is the task of the kidneys (page 142) to remove the surplus water so as to keep the composition of the blood at the correct level. This mechanism would however be greatly strained if a few litres of water were added every time we took a bath or a walk in the rain.

A non-waterproof skin would lead to further trouble by shrinkage on bathing in sea water, which is, in the chemical sense, more concentrated than blood. This problem is faced by fishes which, like ourselves, have blood much less concentrated than sea water. When swimming in the sea, we rely on the impermeability of the skin to keep us from shrivelling. A fish, however, has to process great quantities of water to obtain enough oxygen to keep alive and it is difficult to prevent body water from passing out, across the gills, at the same time as oxygen passes in. The fishes' answer to this problem is explained on page 148.

The waterproofing of the skin resides in the outer layers of the epidermis, that is, in the dead, flattened cells which give the skin its characteristic horny surface. If the outermost cells are removed by peeling a piece of adhesive tape from the skin, water and other substances pass in both directions much more freely. Although the skin is, for practical purposes, quite waterproof, a good deal of water passes through it in two different ways.

The sweat glands (page 46) are tiny coiled structures buried in the dermis, each terminating in a fine tube or pore leading to the outside of the skin. The sweat glands are important in maintaining the body's heat balance; considerable amounts of energy are lost in the form of latent heat by the evaporation of water brought to the surface of the body. There is also an appreciable movement (about 500 grams per day) of water by diffusion through the skin, quite independent of the sweat glands. This water is also evaporated and helps in the removal of heat from the body.

Apart from water, very little of the materials in the body escape through the skin. Sweat is more than 99% water; legends about sweating blood have no foundation in fact, though the apocrine glands (page 134) produce red sweat in the hippopotamus and, very rarely, in man.

The passage of chemicals into the skin from outside is of some medical interest, since this process is a justification for the use of ointments. An ointment is usually an emulsion, containing a drug suspended in an oily or waxy medium. As a way of introducing a drug into the body, smearing the skin with ointment is more effective than merely painting it with a solution — partly because the ointment does not evaporate at all readily and also because it can be applied in greater thickness than a liquid. Most drugs pass through the skin very slowly indeed and it is difficult to obtain a worthwhile concentration by application of ointment.

The effectiveness of the skin in keeping out drugs is emphasized by experience with industrial chemicals. Many of these materials are very toxic but few cases of poisoning by absorption through the skin have been reported. The modern pesticides which are used to an increasing extent in agriculture have been blamed for a number of deaths — but, even here, harmful effects are produced only by gross contamination of skin resulting from neglect of elementary precautions.

Many gases can pass through the skin rather freely. Since, except for oxygen and nitrogen, the air does not normally contain any gases in appreciable amounts, this property is not of great significance — except in biological warfare; mustard gas and more modern products such as sarin can pass through the skin.

The skin takes a very small part in respiration, gaining about 150 ml of oxygen per hour (sometimes more) from the air and losing the same volume of carbon dioxide, which is a waste product from many chemical processes going on inside the body. This exchange represents only about 0.5% of the body's gas transfer, the remaining 99.5% being achieved in the lungs.

The ability to breathe through the skin appears to be a needless luxury, since this method of gas transfer could never be of any practical value. It is in fact a consequence of the economy with which the body has been designed. The outer surface of the body and the inner surfaces of the lungs, digestive tract, bladder and other cavities communicating with the outside, are all composed of epithelial cells. These cells, though they serve many different purposes, have certain basic similarities and it is therefore not surprising that the epidermis should have some capacity for respiratory exchanges.

Fur or hair?

Man has been called the naked ape. The apes, chimpanzees and other primates, which are our nearest neighbours in the biological sense, all have a thick covering of fur, as do most other animals. Fur is an exceptionally good insulator. Eskimo dogs lose no heat from the body even when sleeping in a gale at a temperature of −40°C. At the other extreme, the outside of a sheep's fleece can reach a temperature of almost 100°C in tropical sunlight, while the inside remains cool. Fur gives protection in two different ways. Firstly it is inherently a good insulator. Secondly it traps a layer of still air which is even more effective as an insulator.

Fur is obviously useful to animals living in a cold climate. The human animal developed in warm regions where fur was not needed. It is, however, interesting that the number of hairs per square centimetre of skin is about the same in man as in the other primates; the difference is that human hair (except on the head) does not grow to the same length or thickness as in furry animals.

In engineering terms, the relative merits of fur and smooth skin are related to the problem of maintaining the internal organs of the body at a constant temperature. In a cold climate, the energy needed comes from food and is effectively conserved by thick fur. Sometimes, of course, animals living in cold climates do have to rid themselves of surplus heat − for example during hunting or other violent activity. The fur is such a good insulator that virtually no heat can be lost through it. Arctic animals involved in vigorous exertion at low temperatures can rid themselves of surplus heat in two ways. The legs, which are usually not so thickly coated with fur as the rest of the body, can be stretched out and the counter-current mechanism (page 135) which normally conserves body heat can be temporarily abandoned. Alternatively, heat may be lost by evaporation of water from the mouth.

The problem of removing surplus heat is not confined to arctic regions, for it applies also to thickly-furred animals such as dogs, cats and sheep, even in temperate climates. A dog is not able to rid himself of very much heat by sweating – but on a hot day he pants, apparently breathing in and out rapidly through the open mouth. To the engineer, this seems a singularly inefficient method of cooling, since only a little of the air actually comes in contact with the moist surfaces of the tongue and mouth. The nose contains a much better mechanism for evaporating water and adding it to the air, with consequent loss of body heat (page 95). Merely breathing through the nose would not however produce the desired effect since most of the heat given up to the inspired air is recovered as the air is breathed out again.

Very recently, it has been found that a dog overcomes this problem in a most ingenious way by breathing in through the nose and out through the mouth. On average about three quarters of the air inhaled through the nose is exhaled through the mouth – but the proportion can vary between 0 and 100%. In this way a dog can regulate the amount of heat loss according to the thermal stress imposed by physical activity. Cats and sheep also pant – but whether they use the same ingenious arrangement as the dog is not yet known.

In man, as we have already seen (page 51), surplus heat is removed mainly by the sweat glands. Other animals, including the dog and the pig, are quite well supplied with sweat glands but do not use them to any great extent. The heat control mechanism in man is very well designed. Its effectiveness depends not only on the activity of the sweat glands but on the great number of small blood vessels reaching into the skin, far more than are needed to supply the tissue with oxygen and nutrients. When the need arises, the blood flow in these vessels carries a considerable amount of heat to the surface, for disposal by evaporation of sweat from the skin.

Our adaptation to cold weather is less than perfect; nakedness is pleasant on a hot day but would be fatal in winter. The elaborate design and structure of the skin allows man the ability to work and, more important in early times, to hunt right through the heat of the day. South African bushmen still exploit this ability by hunting in the hottest part of the day, gaining an advantage over other animals which are resting.

Though man has lost his fur in the course of evolutionary development, he still has a good deal of hair. Only a few small areas of the body, including the soles of the feet and the palms of the hands, are completely free of hair. Elsewhere, the skin is quite richly provided with follicles, originating in the epidermis, reaching downwards into the dermis and each housing one hair which, shielded by the walls of the follicle, passes through the skin into the outside world.

Though hair in man (and more especially in woman) is now mainly decorative in function, it does have other uses. Head hair protects the brain against excessive heating from the sun and other hair guards various orifices of the body, including the ears and the nostrils. Eyebrows are said to protect against the glare of the sun and eyelashes have a similar purpose to the well-developed whiskers of the cat by giving warning of the proximity of insects or foreign bodies; indeed, many hair follicles are surrounded by sensory nerves, making the hairs delicate organs of touch.

Hair is proverbially able to stand on end, and the process is not entirely imaginary. Animal skin is well supplied with muscles which, on a cold day, can cause the individual hairs to stand at a greater angle, thereby increasing the thickness of the fur and improving its insulating properties. The same change occurs in response to fright, presumably to increase the apparent size of the animal. In man, vestiges of the mechanisms still remain, as is shown by the gooseflesh associated with cold or fright.

Almost every hair follicle has a side branch leading to a small gland producing sebum, a material rather like ear wax, which makes its way up the hair, to which it gives the characteristic shiny appearance. On parts of the body where the hair is not conspicuous, the sebum spreads out to form a waterproof coating which we remove at frequent intervals by soap and water. By a curious design feature, the female nose is unusually well supplied with sebum and the shiny surface which results needs regular attention from the powder puff.

Hair grows at about the rate of a third of a millimetre per day. On the scalp, about 80% of the hairs are growing at any given time and the other 20% are waiting to fall out or to be removed by the comb. Normally, mature hairs which fall out are replaced by fresh growth. Gradually however the hair follicles of the scalp close up and support no further growth.

Baldness is a process which really begins before birth. The human foetus, at about five months, has a considerable covering of hair on the forehead as well as on the scalp. The hair on the forehead begins to disappear before the baby is born and does not last long thereafter. Much of the fine hair that covers other parts of a new-born baby also disappears to give the proverbial smoothness of skin at the age of a few months. Loss of hair follicles beings to be noticeable again in middle age or sometimes earlier and is an affliction which has no known cure.

Individual hairs (which have not been subjected to heat or chemical treatment) are usually smooth cylinders; in this situation, the hair, in bulk, is more or less straight. In wavy hair the individual fibres are round at some points and oval at others. Where the fibres are ribbon shaped the hair is kinky in appearance.

Heat and cold

The skin, as we have seen, is mainly important as a protective barrier but it has other functions as an organ of the body, notably its sensitivity to heat, cold, touch and pain. Its sensitivity to the last three of these processes is remarkable. We can for example detect a deformation of the skin amounting to only about a hundredth of a millimetre.

Sensitivity to heat and cold is sometimes thought to be rather crude, as evidenced by the well-known experiment in which the observer puts his left hand in a dish of ice cold water and his right hand in a dish of hot water. He is then asked to put one of his hands into a dish of warm water at an intermediate temperature. To the left hand, the warm water appears hot but to the right hand it appears cool. It is also a matter of common observation that the bath water which appears very hot when we first step into it appears to cool quite considerably after a short immersion – even though a thermometer shows little change. The skin does, however, have specialised nerve endings sensitive to quite small alternations of temperature.

Some points on the skin are sensitive to heat, some to cold and others to pressure. If the nerve endings immediately under the sensitive spots are examined (not a painful procedure, since only tiny pieces of skin need to be removed) they are found to be all alike, with nothing to indicate why they respond to different sorts of stimulus.

This is not the only mystery in the performance of the skin as a sensory organ; fire-walking is another. Reliable reports of people walking over glowing embers have been available for more than 2,000 years. The practice is commonly associated with religious or magical ceremonies and is often thought to have some supernatural explanation. Careful experiments made under eminent scientific supervision in London in 1935 and 1937 showed that fire-walking is in no sense a trick and requires no special preparation of the feet or of the subject. Poise, timing and confidence are important; with these qualities, many Europeans have emulated the oriental fire-walkers without injury.

A somewhat similar achievement is attributed to King Edward VII who, as a boy of seventeen, was sent to receive instruction from Lyon Playfair, Professor of Chemistry in Edinburgh. According to legend the future king, on Playfair's instructions, plunged his hand into a cauldron of boiling lead and ladled out some of the liquid metal without sustaining any injury. This particular story has improved with the passage of time but a reconstruction of the true version was made in 1938 by Professor James Kendall, one of Playfair's successors. In this experiment, the molten lead was poured from the cauldron and the professor's daughter passed her hands slowly, with fingers apart, through the stream as it fell.

It is strange that the skin, normally so sensitive to a small change in temperature, can withstand such gross insults without injury. The explanation is not at all clear but it is possible that the grease or moisture normally present in the skin vaporises and forms a protective layer.

The skin is a remarkable structure. It is the largest organ in the body, with a mass of 3 kilograms − twice as much as the brain or the liver − and it contains about 10% of all the water in the body. Well designed for a wide variety of functions, it protects us against the thermal, mechanical and chemical hazards of the external environment. Through its sensitive nerve endings it provides us with information about the environment and helps to guide many of the activities of the body. Through its sweat glands and very efficient network of blood vessels it regulates the body's heat balance and contributes significantly to the water balance.

The many functions of the skin are performed with the economy of design and construction so often found in the human body. Perhaps the most remarkable example of this economy is that the female breast, so effective in both its functional and its decorative roles, is really a modified sweat gland. Clearly the design and performance of the body's outer wrapper are just as impressive as those of its contents.

5: camera or computer?

Medical students, before they are allowed to approach the secrets of their chosen profession, are required to go through an initiation ritual, which involves memorising and reproducing answers to a selection of traditional questions. One such question asks for a comparison between the eye and the camera. The answers vary considerably in scope and content – but seldom display the depth of understanding revealed by a student who once told me: 'The eye and the camera are entirely different. The camera is made of tin but the eye is made of meat'.

Roughly speaking, the eye does behave like a camera, or perhaps a cine camera, since it deals quite well with moving objects but does not store up a pile of snapshots to be viewed at leisure when there is nothing else of interest in sight. A photographic dealer would however view the eye with somewhat limited enthusiasm, because the design and manufacture have rather conspicuous defects.

To appreciate the merits and shortcomings of the eye, we should first examine the design and operation of the camera. The essential requirements of a camera are as follows.

(a) A light-sensitive film or plate on which is formed an image of the object or scene which the photographer wishes to record.

(b) A lens to form the image. The single lens used in simple cameras gives an image which, though acceptable for many purposes, has a number of defects. The most obvious is that, if the centre of the picture is in sharp focus, the outside is somewhat blurred; this fault is known as technically spherical aberration.

Another cause of unsharpness is that the focusing power of the lens is different for different colours – an effect known as chromatic aberration. These defects can be overcome by using a lens with several components.

(c) A means of focusing the lens. To obtain a sharp image, the distance of the lens from the film must be greater for a closeup than for a landscape. In a simple camera, the lens is mounted in a threaded tube with a scale around its lip indicating the amount of rotation needed to secure sharp focus for objects at various distances. Sometimes the distance is estimated by eye. Alternatively, the camera may be fitted with a rangefinder – an optical device in which, when two images are made to coincide, the distance of the object is

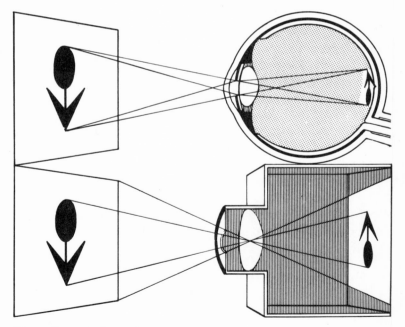

Figure 3 eye and camera

indicated on a scale. A further refinement is the coupled rangefinder in which the adjusting knob is linked by levers or cams to the lens so that, when the rangefinder is in correct adjustment, the lens is accurately focused.

(d)   Means of adjusting the amount of light reaching the film. The production of a satisfactory photograph requires a certain quantity of light, depending on the sensitivity (or speed) of the film. The amount of light reaching the film may be controlled by a shutter which normally keeps the inside of the camera completely dark but, under the action of a spring, can be opened for a fraction of a second. In simple cameras, a shutter is enough but the photographer has greater scope if a diaphragm is fitted. The diaphragm consists of an assembly of overlapping metal wafers shaped so as to leave a central circular aperture which can be adjusted in diameter by a ring mounted on the lens barrel. On a dull day the whole of the lens may be needed but in bright light the outer parts can be blacked out, leaving a smaller aperture for the passage of light.

In a simple camera, the choice of shutter speed is often made by rule of thumb. For more reliable results, reference may be made to an exposure meter — a small photo-electric cell coupled to a meter indicating the correct exposure for a particular lens aperture and film speed. Sometimes the current from the photo cell (or from auxiliary batteries) is used to adjust the lens aperture and the shutter control.

It might seem superfluous to have both a shutter and a diaphragm since either is capable of regulating the amount of light reaching the film. Consequently a large aperture and short exposure will transmit the same amount of light as a small aperture with a long exposure. The combination of diaphragm and shutter does however allow the photographer to obtain satisfactory results over a wide range of light intensities and gives him additional help in certain special circumstances. The lens aperture determines the depth of focus, which is often important in photography. With a small aperture, objects will be rendered in sharp focus even though some are relatively near the camera and others far away. With a large aperture, the focusing is much more critical. For portrait photography, it may be desirable to use a large aperture so that the background will be out of focus. A short exposure and correspondingly wide aperture is also necessary for the photography of moving objects. On the other hand the recording of architectural detail is best achieved by a small aperture and relatively long exposure. Another limiting factor arises from the fact that the average photographer cannot hold the camera absolutely steady for exposures longer than 1/50th of a second. Consequently unless a tripod is used it is better to opt for a short exposure.

The photographer's exposure meter was at one time a relatively bulky object and was held in the hand. As miniaturization developed, it became possible to build the exposure meter into the camera, usually occupying a position above the lens. Unfortunately the exposure meter does not, in either of these arrangements, see exactly the same scene as the camera lens and its indications may therefore be inaccurate. A further refinement involved the construction of an exposure meter with its light-gathering element in the shape of a ring around the lens. More recently the problem has been satisfactorily solved by the development of through-the-lens exposure meters. Here, a mirror is used to intercept light which has passed through the camera lens and reflected, with the help of auxiliary lenses and prisms, into a view-finder. The light going into the photographer's eye is sampled by two photo-electric cells (built into the viewfinder) which transmit the appropriate exposure and aperture settings. The control which operates the shutter also moves the mirror out of the way so that light which has passed through the lens then reaches the film directly.

Here again we see technology imitating nature. The eye's exposure meter, controlled by the visual computer in the brain and the retina, uses signals from light which has passed through the lens, though without the need for mirrors and prisms.

Moving pictures

In basic outline, the eye first shows a lens which forms an image. Instead of film there is the retina, a kind of screen from which signals are transmitted to the brain. Confronted by a blueprint of the eye, the camera enthusiast would immediately complain that a great amount of complicated mechanism is permanently glued onto the film. The retina is not, as might be expected, placed directly in the path of the light coming through the lens. On the contrary, it lies at the bottom of a complicated pile of light-sensitive cells and other tissues, surmounted by the fibres of the optic nerve. In other words, an outpost of the brain is spread over the retina. With all these obstacles in the way, how do we see an image at all?

The answer to this question involves one of the most striking properties of the eye, its ability to concentrate on movement and to reject stationary images almost as soon as they are formed. This effect is not easily demonstrated, because it is almost impossible to keep the eye absolutely still for more than a fraction of a second. This restlessness may be studied by placing a contact lens on the eye, glueing a tiny mirror to the lens and reflecting a beam of light from the mirror onto a moving film. Experiments of this kind show that the eye is never still. A rapid tremor goes on all the time. When reading or looking at a picture, the eye makes small jerky movements, with resting intervals of about 1/10th second. When following a moving object, the eye moves more smoothly.

It seems that the satisfactory operation of the eye does not need a stationary image on the retina. Using a contact lens with a mirror it is not difficult to compensate for the natural movement of the eye, thereby ensuring that the image on the retina is absolutely stationary. In these circumstances, the object being viewed is at first seen quite clearly but, after one to two seconds, becomes quite faint. After three or four seconds more, the eye records only a uniform greyness with no detail. After a further short interval, the field of vision may become quite black and nothing at all is seen. Alternatively a faint image of the original object may appear and disappear; in this situation, a test object such as a circle or a square will sometimes be partially seen, as one or two lines. By adjusting the external mirrors and lenses used in this experiment, the image on the retina can be moved through a small distance. When this is done, good vision is restored but only for a few seconds.

It seems therefore that the eye is immune from camera shake, the bane of the amateur photographer and, indeed, cannot maintain a sharp image unless the eye is continually moving.

Eye shadow

Since the eye (or, more precisely, the combination of eye and brain) responds to the movement, rather than the mere presence, of an image on the retina, the permanent shadows of the nerve cells, blood vessels and other structures lying close to the retina are not noticed. It might however be expected that these structures could be detected by their effect on the image moving from one part of the retina to another, in response to the incessant movements of the eye which we have already noted. It is possible, by careful experiments, to produce on the retina perceptible shadows of some of the anatomical structures. In practice, we seldom have any trouble from this cause, because the fibres of the optic nerve are mainly concentrated near the outside of the retina, by-passing the central area where the most acute vision is provided.

When choosing a camera, the photographer will take note of its f-number, expressing the maximum aperture of the lens (on which the light-gathering power depends) as a proportion of the focal length, which is roughly the distance between the lens and the film. A large aperture lens will obviously allow a picture to be taken in a shorter time and is therefore to be preferred. The focal length is usually denoted f and the lens of a cheap box camera might be classified as f/16, meaning that the aperture of the lens is 1/16th of its focal length. A more expensive lens might have an aperture of f/4 or even f/2 which is about the limit for all but the most costly cameras. On this scale, the eye lens has quite a large aperture, corresponding to f/2 or slightly better.

The photographer does not always use the maximum aperture, partly because he usually obtains a sharper image by using only the central portion of the lens, and also because, in bright sunlight, he has to reduce the aperture to avoid over exposure of the film. Control of the aperture of the lens is achieved by an iris diaphragm defining the area of the lens through which light is allowed to reach the film.

The diaphragm of a camera imitates the iris of the eye, easily recognised because it gives the characteristic blue or brown colour. The automatic diaphragm of a sophisticated camera copies what is done without effort in nature. Stand in front of the mirror; put a hand over one eye and notice how the iris of the other eye opens to let in more light. This simple experiment illustrates the automatic adjustment, but does not tell the whole story. The eye can, with little difficulty, adapt to light intensities over a range of 10,000-1 – but the maximum and minimum areas defined by the iris are in the ratio of about 16-1. It seems that the sensitivity of the eye is related to the brightness so as to maintain an output signal more nearly constant than the widely fluctuating light input, a facility which may be compared with the automatic volume control used in radio receivers.

Light and dark

When a photographer wants to work in dim light, he changes to a faster film which, though more sensitive, gives a grainier picture in which fine detail is not so clearly rendered. The eye does something rather similar. The structures in the retina providing the initial response to light are tiny cells of two kinds, the cones (which operate in bright light) and the rods, which come into action in dim light. Cone vision allows good detail because each cone is usually connected to a single fibre in the optic nerve which transmits signals to the visual computer in the brain. Consequently the requirement for visual resolution of two nearby objects, or two parts of a larger object, is that their images on the retina should be far enough apart to fall on two different cones. In practice, this corresponds to a separation of about an inch at a range of 200 yards, though, as we shall see later, the eye can often do a lot better than this. The rods, on the other hand, do not operate a single unit but are grouped in bundles for connection to the optic nerve. Consequently it is much harder to make out small objects or to observe fine detail in dim light. Cones are capable of colour vision but rods are not; that is why, as the proverb tells us, all cats are grey at night.

A production engineer might object to the eye because the quality control is not very strict. In a camera, the focal length of the lens is fixed and, when the focus control is set for infinity, the distance between lens and film must be accurately equal to the focal length if a sharp image is to be obtained. In the eye, the focal length of the combination represented by the cornea (a tough curved outer layer which provides most of the bending power for arriving light) and the lens varies from one person to another with a range between 14 and 18 mm. The front-to-back length of the eye, naturally somewhat greater, varies between 22 and 26 mm. Variations in the dimensions of lens and cornea and in the distance between the lens and the retina are however not entirely random. In any individual person they are related to one another in such a way that most of the population have reasonably normal eyesight.

Far-sighted

Careful measurement does show that most people, about two

thirds of the British and American population, are long-sighted, that is, better able to see objects at a distance than those close to the eye. The departure from perfect vision is generally quite small and can be accommodated without the need for spectacles — but the difference has left its mark on society, as shown by the preference for entertainments, such as football and the cinema, requiring good distant vision.

In all but the cheapest cameras, the lens is moved in accordance with the distance of the object to make sure that a sharp image is always produced on the film. The eye does much better, for its lens automatically adjusts to the distance of the object; in this respect, the eye lens is even more sophisticated than the zoom lens now favoured by film and television camera men. In the zoom lens, the focal length is changed by moving one or more components of the lens, but in the eye the focal length is adjusted by changing the shape of the lens. A relatively flat lens has a long focal length and one which bulges more noticeably at the centre as a short focal length. Glass is not a good material for instant adjustments of this kind, but the soft squashy material which makes up most of the body is much more versatile. The lens of the eye lies inside a ring of muscle (called the ciliary muscle) to which it is attached by a network of elastic fibres. If the ciliary muscle contracts, the tension in the fibres is reduced and the lens relaxes to a more convex shape, that is, with a more distinct bulge at the centre. On the other hand, when the ciliary muscle stretches, a greater tension is transmitted by the fibres holding the lens, which accordingly takes up a flatter shape with a longer focal length.

Non-stop growth

This mechanism works very well in most people until they reach middle age. After that, the adjusting power of the lens — usually known as accommodation — begins to fail for a number of reasons. For one thing, the lens of the eye has the unique property that it never stops growing. Perhaps this is not absolutely unique, for the same could be said about hair and skin. However, dead skin is rubbed off every time we shake hands or open a door and surplus hair can be removed with scissors. The eye lens grows from the centre and adds fresh cells continually. The older cells near the centre are increasingly cut off from the blood vessels delivering oxygen and other nutrient materials. When these cells eventually die of starvation they cannot be removed. Consequently the lens continues to grow throughout life. This increase in size is one reason for the loss of accommodation characteristic of middle and old age. Imagine a disc suspended at the centre of a hoop by spokes made from elastic bands. If the disc is small, the elastic bands will be stretched tightly, but if the disc grows bigger, the elastic will slacken. A rather similar effect occurs in the eye. Though the lens grows throughout life, the ciliary muscle and the rest of the tissues stop growing, along with the rest of the body, when adult age is reached. Consequently, as the lens continues to grow, the elastic ligaments joining it to the ciliary muscle become slacker and therefore less able to control the shape of the lens. The progressive hardening attributable to the dead cells near the centre of the lens also makes control of its shape more difficult.

A photographer, knowing that a simple piece of curved glass cannot give a perfect image, usually finds it worthwhile to pay a lot of money for a lens which is fully corrected. The design and performance of the eye are not up to the highest standard in this respect. The first fault of a simple lens is a distortion of the image technically known as spherical aberration. This defect arises because rays of light near the axis of the lens come to a focus further away than those from the outer parts of the object. In the eye this problem is solved in an ingenious way. The cornea, which provides most of the refracting power, is not quite spherical but is more sharply curved near the centre than at the edges; the refractive index (or bending power) is also greater near the centre. For these two reasons, rays of light near the central axis are more distinctly bent so that they come to focus at about the same place as rays passing through the outer parts of the lens, and the effect of spherical aberration is almost eliminated. aberration is almost eliminated.

The next problem for the camera designer is chromatic aberration, which causes blue light to be brought to a focus closer to the lens than red light. The usual solution is to use a combination of lenses made from different kinds of glass. Unfortunately the range of optical materials available in the eye is not enough to provide a colour-corrected lens.

The painter's eye

The difficulty is tackled in three ways in the human eye. The lens is not absolutely transparent but is slightly yellow in colour and stops light in the violet and ultra-violet regions of the spectrum, where chromatic aberration would be greatest. People who have the eye lens removed for cataract and replaced by clear glass can see quite well by ultra-violet light. The colour-correcting action of the lens is reinforced by a patch of yellow pigment which covers the central part of the retina, used particularly for detailed vision in good light.

The built-in filter is a device quite familiar to photographers, but in the eye its action becomes more pronounced as we grow older. Some scholars believe that the preponderance of orange and reddish tones in Turner's later paintings are attributable to this effect.

The eye has a third method of dealing with chromatic aberration. The cones, used for seeing in bright light, are more sensitive to colours near the red end of the spectrum than to blue or violet. In dim light, colour vision is not good and chromatic aberration is not much of a problem. For acceptable rendering of colours in bright light it is necessary to discard the part of the spectrum in which chromatic aberration would be pronounced. The yellow colour of the lens, the yellow filter over the centre of the retina, and the difference in colour sensitivity between the cones and the rods all act in the same direction for this purpose.

Apart from spherical and chromatic aberrations, the eye usually displays astigmatism. Because of uneven curvature of the cornea, vertical and horizontal lines are not rendered with equal sharpness in their images on the retina. This defect is easily corrected by spectacles with a compensating curvature. However, an astigmatic painter, working without his spectacles, might find vertical lines sharply focused and horizontal lines somewhat blurred (or vice versa). He would, in his painting, emphasize the lines that he saw more clearly.

It has been suggested that the elongated figures of El Greco's paintings reflect his astigmatism – but this speculation is of doubtful validity. X-ray examination of some of his pictures shows that the original outlines were of normal proportions, the distinctive elongation having been added at a later stage in painting. There is also a more fundamental objection. The artist paints what he sees. If the viewing system (composed of eye and brain) makes an elongated image of a sitter or a scene, the painter's representation on the canvas must be an exact likeness of the original so that it too, when processed by his eye and brain, will produce an elongated image.

Fast or slow?

The keen photographer uses more than one kind of film. When working at night or indoors he usually prefers a fast film, with good sensitivity to light but not so good at reproducing fine detail. When more light is available, he will prefer a slower film with a fine-grained structure giving a better quality of picture. In the eye, the sensitivity to light and the ability to detect fine detail are adjusted automatically according to the prevailing conditions. The change in sensitivity takes a little longer than the removal and replacement of a roll of film. After changing from bright light to near darkness, a period of twenty or thirty minutes is needed for the eye to become completely adapted. It might be thought that this rather long delay would have inconvenienced primitive man, surrounded by other creatures often possessing very good night vision. However, the interval of time between daylight and darkness, even in tropical latitudes, does not fall below half an hour and our ancestors therefore had enough time to adjust their defences as daylight faded. The dark adapted eye has a sensitivity which even now cannot easily be matched by any man-made photoelectric device; certainly no electronic instrument of comparable size, convenience and general availability has a performance remotely approaching that of the eye in dim light. The sensitivity of the eye is indeed very close to the ultimate theoretical limit imposed by the quantum theory of radiation. A quantum is the smallest amount of a particular radiation that can exist. In the blue-green region, where the eye is most sensitive, a visual signal corresponding to something between 50 and 150 quanta striking the outside of the eye (that is, the cornea) will produce the sensation of seeing. The sensitivity of the retina is even better than this figure indicates, because about half of the quanta are lost, by absorption and scattering, in passing through the cornea, the lens and other components of the eye. This leaves 25-75 quanta on arrival at the retina. Only about a fifth of the light reaching the retina is absorbed in the rods and therefore available for visual process. We can therefore conclude that the threshold of seeing corresponds to the absorption of 5-15 quanta of light in the rods, which are the sensitive cells of the retina for night vision.

The sensation of seeing is triggered off by chemical changes in the retina. The sensitive material for this purpose is rhodopsin, sometimes known as visual purple. Rhodopsin is actually a compound of retinene (a material somewhat similar to vitamin A) and opsin, which is a protein. All proteins have very large molecules and the opsin molecule has, on its surface, a cavity into which a retinene molecule fits neatly. When the energy of a quantum of light is absorbed, the retinene molecule changes its shape and springs out of its niche. The chain of events which follows is not yet fully understood but the eventual outcome is that a nearby nerve cell generates an electrical signal which eventually passes to the brain − although, to be precise, the retina is really an outpost of the brain and not a separate organ.

Threshold of vision

Since the visual process is started off by the absorption of a single quantum in a single rod of the retina, it might be thought that the threshold of vision corresponds to a single quantum; this however is not the case. The minimum threshold for vision is obtained when the light delivered to the retina is concentrated into an interval of no more than a 1/10th second and spread over an area of no more than about 500 rods. If the same amount of energy is delivered in a flash of longer duration or is spread over a larger area, the sensation of vision will not be produced. Experiments of this kind indicate that the threshold of vision depends on the distribution of the incident light, both in time and in space, and not merely on the total amount of energy delivered. It seems clear that absorption of one quantum can produce the chemical effect and presumably the accompanying electrical impulse in a fibre of the optic nerve − but the brain requires rather more evidence to be convinced that something has been seen.

In discussing the ultimate sensitivity of the eye, we observed that four fifths of the light reaching the retina passes straight through without being absorbed in any useful way. This situation is well known to photographers, since only a fraction of the incident light is absorbed in the emulsion of a film. There is a danger that some of the light which has not been used in passing through the sensitive emulsion may be reflected or scattered by the celluloid base and may find its way back into the sensitive layers, where it will merely cause confusion in the form of general fogging. Modern photographic film has a dark coloured backing to trap this light and the eye is equipped with a very similar safety device. The back of the retina contains a layer of cells containing a dark brown pigment known as melanin, which is also responsible for the colour of sunburnt skin. The melanin layer effectively removes from the scene all the light which has not been usefully absorbed in its first passage through the rods and cones. Animals which are active during darkness use a different mechanism, consisting of a reflecting layer which sends the unused light back into the sensitive part of the retina to reinforce the image. There is some loss of definition in this process, but many animals do not have very sharp vision to start with and the gain in sensitivity is important for night vision. The reflecting layer at the back of the retina is responsible for the luminous appearance given by the eyes of cats or other nocturnal animals when seen, for example, in the beam from a car headlamp.

Black or white?

The eye offers much for the engineer or the photographer to admire. As we have seen, many features of its performance can be explained only by using sophisticated ideas of science and engineering. Some of the achievements of the eye are even now very difficult to understand. An old proverb observes that 'black will take no other hue'. But why not? The colour of an object depends on the light reflected from it into the eye. A piece of chalk looks white because it reflects most of the incident light, quite impartially as to colour. Soot looks black because it reflects rather little of the incident light. But a streak of soot viewed in full sunlight reflects much more light than a piece of chalk viewed on a dull day. The colours of these and other familiar objects remain constant because the sensitivity of the eye is automatically adjusted in accordance with the average illumination over the whole field of vision. The iris plays a very small part in this process which is controlled to a much greater extent by interchanges between the eye and the brain — though the details of this mechanism are not yet understood.

The eye is much more than a photo-cell, merely converting visual signals into electrical impulses. Animal experiments, involving the direct recording of electrical signals from fibres of the optic nerve or from parts of the retina, suggest that the eye has mechanisms for signalling to the brain when the intensity of illumination is increased or decreased or when an object in the field of view is moving. Other cells generate continuous signals so long as the light pattern reaching the retina remains unaltered, but become silent as soon as there is any change in the scene, rather like a sentry who keeps firing his rifle to show that nothing much is happening.

Though the eye is impressive as an optical instrument, it sometimes appears to do even better than its design allows. The diameter of a single cone in the retina is about 0.0015 mm. To produce an image which will just fill one cone, an object should subtend an angle of about half a minute at the eye; in angular measure a minute is a sixtieth of a degree and half a minute is the angle corresponding to an object one inch across at a distance of 200 yards. It might reasonably be thought that an image covering one cone represented the absolute limit of resolving power for the eye. It is however easily shown that the eye can clearly see a distant telegraph wire which subtends an angle of as little as half a second — that is, sixty times less than the theoretical limit. There is really no satisfactory explanation for this achievement but it may be related to the fact that the eye is always moving. For this reason, the image of a long thin wire will fall at different places on the retina, slightly affecting a large number of cones. The information gathered in this way is processed in the retina or in the brain to form the image of the wire.

In general the eye is very good at detecting boundaries but does not seem to care quite as much about featureless areas of uniform illumination. It is sometimes asserted that the eye produces automatic contrast improvement (such as is sometimes practised by photographers) to enhance the difference between adjacent light and arc areas. Information supporting this view has been obtained by direct recording of signals from the optic nerve of the horseshoe crab, where each visual cell is permanently connected to its own nerve fibre. In this system, bright illumination of visual cells leads to a reduction in sensitivity of neighbouring cells, thereby enhancing the contrast between light and dark areas. The existence of this effect in the human eye is however still uncertain.

The simple notion that the retina is merely a screen corresponding to the film in a camera, and that the brain somehow sees the image impressed on it, would not disgrace a first year medical student, but the true story is much more interesting. The retina is merely an extension of the brain and is very much a three dimensional structure with its own elaborate data processing facilities. The eye is an exceptionally sensitive optical instrument displaying many striking features of design and performance; even the windscreen washers and wipers have not been forgotten.

6: well-tuned sounds

Hearing and balancing

Sound is a form of motion. Music is made by the vibration of air columns, strings, stretched skins or reeds, all of which launch invisible waves which travel through the air and set up corresponding vibrations in the piece of skin which forms the ear drum. Hearing is the sense by which the brain perceives and interprets the disturbance produced by the vibrating object — but it is much more than a distant sense of touch.

The sensitivity of the ear is remarkable, coming close to the theoretical limits imposed by the structure of matter and the basic laws of physics. As we shall find later in this chapter, the sensation of sound can be produced by a movement of the ear drum of a thousand millionth of a centimetre, which is considerably less than the diameter of a hydrogen atom. The ear also displays a judgment of pitch which is unequalled in the animal kingdom and makes man the most musical of all living animals.

Apart from its refined sense of hearing, the ear is also an organ of balance. A four-legged animal is inherently stable because its centre of gravity, for all normal attitudes and postures, falls well within the rectangle defined by the points where the feet touch the ground. For a two-legged animal such as man the equilibrium is more precarious. A damp footmark in the bathroom shows that the area in contact with the floor is only a few square inches — but the circumference of the body may well be forty inches at the hips and not much less at the chest. If a vertical drawn through the centre of gravity lands outside the small area where the feet make contact with the ground, the body falls down. A life-size model of a man is easily toppled and is more likely to fall over than to remain upright, not unreasonably in view of the shape, with a broad top (the trunk) tapering to a narrow base at the feet. Without the elaborate control and stabilising mechanisms provided in the ear, we should find it very difficult to remain upright or to walk about, and impossible to run up and down stairs or perform other commonplace but (in engineering terms) demanding feats of acrobatics.

The important functions of the ear, hearing and balancing, are conducted in quite a small space hollowed out of the temporal bone, which is part of the skull. The protection provided in this way is quite important since the mechanisms of the ear are fragile and could not safely be exposed to all the hazards of the external environment.

How do we hear? An electronic engineer interested in acoustical problems would first arrange a microphone to collect the sounds which he wished to examine, and to convert them into electrical impulses, which are more convenient for recording and subsequent study. The ear does a rather similar job by collecting sounds and changing them into nervous impulses, which are conducted to the brain, where an elaborate data processing system converts them into the sensations of loudness, pitch and quality, and also gives some information about the direction of the original sound.

The structure which carries out these functions has three main parts: the outer ear, the middle ear and the inner ear. The outer ear (all that we normally see) comprises the pinna, a corrugated flap of skin and cartilage leading to the meatus, a narrow tunnel which is a dead end, terminated by the ear drum. The pinna, like the ear trumpet sometimes used by deaf people, helps to catch some of the sound which would otherwise pass by, and to guide it towards the ear drum. The visible parts of the ear are not particularly graceful, but the apparently careless design does have a purpose. The meatus and ear drum would, left to themselves, form a resonant system (rather like a sea-shell) and would greatly emphasise some sounds at the expense of others. The irregular surface of the pinna prevents any sustained build up of this kind.

The action of the built-in ear trumpet, provided by the pinna, is very effective in the cat, where it gives a tenfold increase in the amount of sound delivered to the ear drum. The cat, like many other animals, has a well-developed pinna, controlled by muscles which move it quickly from the relaxed state to the action position. The horse has an intricate system of seventeen muscles to adjust the position of the ear, or, by a twitch, to dislodge flies. In man, the need for acute hearing is no longer of vital importance and only a few gifted individuals have the ability to wag their ears. There are however still remnants of nine muscles around the pinna to remind us that the ability to detect and localise strange noises was often a matter of life and death to our remote ancestors.

Data processing

So far we have seen how the outer ear samples the sounds which are all around us. These sounds are mechanical vibrations; their conversion into signals suitable for processing by the brain brings interesting and difficult engineering problems. The major difficulty is that the body is very different in structure and composition from the environment in which it lives. Sometimes the differences are not of great consequence. Materials can usually be transferred easily from the surroundings to the inside of the body. Oxygen, for example, passes through tissue boundaries in the lungs and then dissolves in the blood which transports it to other parts of the body. The chemical constituents of foods, dissolved by the saliva or the gastric juices, pass into the blood through the lining of the stomach or of the gut. The transmission of sound is more difficult.

The inner ear, where sound vibrations are converted into nervous impulses for further processing in the brain, is, like most other parts of the body, composed of liquid and soft tissues. Although oxygen and foodstuffs can be passed from the air to the blood without much loss, sound vibrations behave in a very different way. In particular, sound travelling in air is almost completely reflected when it strikes the surface of water; a man swimming with his head submerged will not hear a pistol fired in the air immediately above him. Isaak Walton was unduly cautious when, in *The Compleat Angler* he advised his readers 'to be patient and forbear swearing, lest they be heard and catch no fish'.

The poor transmission of sound from air to water, which prevents the fish from hearing bad language (and would, if not corrected, make man's hearing much less impressive), is an example of the problem well known to electrical engineers under the name of impedance matching, and arises mainly from the fact that water is so much denser than air. The

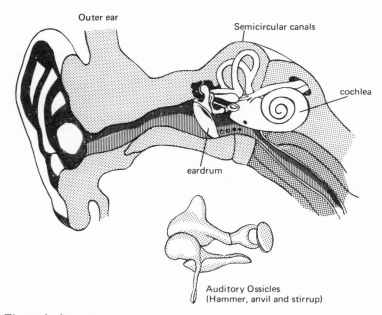

Figure 4 the ear

problem arises in the following way. Sound vibrations are rapid alternations of pressure. In air the pressure changes are relatively small and the amplitude, that is, the distance actually moved by the air molecules as a result of the vibrations, can be relatively large. A denser substance such as water (or any other biological fluid) cannot be compressed nearly as easily. For this reason much greater pressures must be applied to transmit sound vibrations; at the same time, the to and fro movement of the molecules caught up in the vibrations is much smaller than in air.

If a sound wave travelling in air reaches a solid or liquid surface, most of it is reflected and only a small percentage is transmitted. The design of the ear allows this difficulty to be largely overcome. The necessary adjustments are brought about in the middle ear, an air-filled cavity bounded at one side by the ear drum and at the other side by another piece of stretched skin known as the oval window, which leads to the inner ear. The ear drum is connected to the oval window by a chain of three small bones usually known (from their shapes) as the hammer, the anvil and the stirrup; the hammer is firmly joined to the ear drum and the stirrup bears on the oval window of the inner ear.

The ear drum and the three small bones operate as a lever, with an action which may be compared to that of the jack used to lift a car when changing the wheel. Here, the force of a few pounds (supplied by the pressure of the hand on the end of the lever) is transformed into the force of half a ton or more to lift the car. Corresponding to the increase in force is a reduction in amplitude, since the hand may move repeatedly through a distance of a foot or more while the car rises a fraction of an inch. The leverage provided by the ear drum and the three small bones to which it is joined reduces the amplitude of sound vibrations between the ear drum and the oval window by a factor of about fifty, and, of course, increases the pressure associated with them in the same ratio. As a result, sound is passed from the outside air to the liquid filling of the inner ear without much loss.

Harp or telephone?

Vibrations in the air, suitably transformed, have now been delivered to the inner ear. The end product is the complicated sensation of sound registered by the brain. What takes place in between has been the subject of great discussion and controversy during the last century. The arguments may be summarised by asking whether the ear is a telephone or a harp.

The telephone theory supposes that the sound vibrations are converted, in the ear, to electrical (or perhaps chemical) signals, which travel along the auditory nerve into the brain in much the same way as sounds spoken into the mouthpiece of a telephone are changed into electric currents which flow along wires to the exchange. This theory is quite attractive because it reduces the function of the ear to a rather simple level, leaving the complicated processes of analysing and interpreting the sound to be conducted in the brain.

The theory does not, however, stand up to close inspection in the light of what we now know about the nerves and the brain. A current flowing in a telephone line corresponds in two ways to the sound which provokes it. Firstly, the current alternates at a frequency of hundreds or thousands of cycles per second,* to match the varying frequency of the sound. Secondly the amplitude (that is, the size) of the current rises and falls according to the loudness of the sound.

Messages conducted along a nerve do not behave in either of these two ways. When a continuous stimulus is applied to a nerve, the response is not continuous but is a succession of short impulses — more like a Morse signal than a steady current. If the stimulus is made stronger, the resulting impulses are still of the same size, but occur more frequently. After each impulse, the nerve is out of action for about a thousandth of a second and does not respond to any further stimulus during this time. Obviously the auditory nerve, even

---

* The unit of frequency is the Hertz (Hz) which is one cycle per second.

if it behaved like a telephone cable, could not respond to sounds at a frequency greater than 1,000 Hz. Yet the average human ear can cope with frequencies up to 10,000 Hz. The telephone analogy clearly fails.

To make a better theory of the hearing process, we must look at the design of the inner ear in more detail. The inner ear contains structures concerned with the sense of balance but for our present purpose the major part of it is the cochlea, named after the Latin word for a snail. The cochlea is a twisted tube, resembling the shell of a snail or the end of an old-fashioned cream horn. The interior of the coiled tube is divided in two by a thin tough sheet of tissue called the basilar membrane which runs almost all the way to the tip. The basilar membrane, when examined under the microscope, is seen to carry thousands of fine fibrous structures.

In the resonance theory of hearing, proposed by Helmholtz in 1863, each fibre is supposed to have its own resonance frequency and to be set in motion when the appropriate sound arrives. The inner ear, in other words, is rather like a harp with its strings ready to sing when suitably stimulated. It is then not too difficult to accept that the sensation developed in the brain will depend on the place on the basilar membrane from which the signal has come. In this way the brain could respond to sounds of quite high frequency, even though, as we have seen, direct transmission of the corresponding electrical signals along a nerve is impossible.

There are however serious objections to the resonance theory of hearing. For one thing, the fibres in the basilar membrane are not free to vibrate separately like the strings of a harp. They are matted together, more like the fibres of a carpet.

If the fibres in the basilar membrane were indeed to be capable of vibrating in response to the normal range of sounds, some of them would have to be very tightly stretched. This is because, other things being equal, the pitch at which a string vibrates depends on the tension to which it is stretched. The strings of a double bass, for example, are relatively slack and can be displaced without much difficulty by the finger — but the strings of a violin, producing sounds of much higher pitch, are taut. If they were not matted together, the fibres in the basilar membrane could probably withstand the stretching forces corresponding to high pitched vibrations. However it is known from experiments on living or recently dead animals that the basilar membrane, when slit by a knife, does not gape as it certainly would if stretched by such large forces.

An alternative theory, due to von Bekesy, retains part of the resonance theory by allowing that the response to high notes and to low notes will be concentrated at different parts of the basilar membrane. The mechanism is not resonance, but an effect which can be demonstrated at the fireside. If a hearth rug is grasped at one end and shaken gently, a travelling wave moves along it. If the shaking is done slowly, the peak of the wave will be seen halfway or further along the rug. On the other hand if the up and down movement is made more quickly, the hump of the wave will be most noticeable nearer to the end where the shaking starts.

According to present views, the arrival of sound at the oval window of the cochlea causes vibration in the watery fluid within. The to and fro movement of the fluid causes waves to travel along the basilar membrane which (like a carpet) has a large number of fibres standing up from it. These fibres, known as hair cells, are in contact with nerve fibres which eventually run into the auditory nerve.

It is possible, by sensitive electronic techniques, to pick up signals travelling in nerve fibres. Experiments of this kind show that the part of the basilar membrane nearest to the oval window responds to high frequencies and the more distant parts to low frequencies. The signal reaching the brain is therefore coded, according to the position along the basilar membrane of the nerve fibres where it originated. Consequently it is not difficult to hear high and low notes at the same time. If the ear receives a pure tone of feeble loudness, only a small number of nerve fibres are activated, at the appropriate position along the basilar membrane. If the sound is increased in loudness, more and more of the neighbouring nerve fibres are brought into play. The message to the brain therefore contains information about the pitch of the sound and about its loudness. So the ear behaves not exactly like a harp but more like a viola d'amore; the attractive tone of this instrument depends upon the action of a number of sympathetic strings, which come into vibration when a loud note is sounded on one of the main strings.

Hi fi stereo

Having reviewed the structure and basic operation of the ear, we may now examine its performance in more detail. A basic question comes to mind immediately. The ear is a sensitive and useful organ but it is not, like the heart, absolutely essential to life. So why do we need two ears when we have only one heart and one brain? The answer is that the two ears form a stereo system, allowing us to identify the place from which a sound is coming and to tell whether the source is moving. This is an ability more helpful to animals but it was once vitally important to man and is still useful.

Sound travelling from a point straight ahead (or directly behind) will reach both ears at the same time. From any other position, there will be a difference in the time of arrival at the two ears. The maximum effect occurs when the sound comes from a point on an imaginary line running through the two ears, that is, from a point to the left or to the right of the observer. In this case, the time difference is about seven hundred microseconds. This is more than enough for accurate processing in the ear and brain, because experiments have shown that clicks presented first to one ear and then to the other can be distinguished as separate sounds when the time interval between them is as small as 30 microseconds.

It might be thought that the localisation of sound could be done by a simpler method. Sound coming from the left of the head should for example produce more effect on the left ear than on the right. In practice this is true only for high frequencies, more than about 5,000 Hz, or a little beyond the top note on a piano. At frequencies below 2,000 Hz (the top note of an organ) the wavelength of sound is greater than the diameter of the head (17 cm), and sound travels round the obstacle quite successfully; in other words, at low frequencies the head casts no shadow for sound rays and the signal is equally loud in both ears. In practice, sound localisation for frequencies above about 5,000 Hz is based on the difference in signal strength at the two ears. For frequencies below about 2,000 Hz, the difference in time of arrival at the two ears is exploited. In the range between 2,000 and 5,000 Hz, neither of these processes is entirely effective and localisation cannot be performed so accurately. There are of course simple ways of helping the sound location process. The head is instinctively turned so as to produce the maximum time or intensity difference at the two ears. In daylight the eyes are also used to locate an unknown source of sound.

Apart from its direction, the properties of a sound which are important to the listener are loudness, pitch and quality. The loudness of a sound depends on its intensity (that is, the amount of energy which it carries) and on its frequency; the sensitivity of the ear is best for frequencies between 1,000 and 4,000 Hz, and falls off at the two ends of the scale. The human ear does, however, respond to sound at frequencies between 15 and 15,000 Hz. Below 15 Hz separate impulses are registered and above 15,000 Hz most people do not hear at all — though children can often reach a good deal further. The pitch of a sound is its position on the scale and is related, as we have already seen, to the place at which it produces the maximum vibration on the basilar membrane of the inner ear. Pitch depends mainly on frequency but sometimes also on intensity. At frequencies less than 1,000 Hz loud sounds appear to be of lower pitch. Above 3,000 Hz increase of intensity brings a rise in pitch.

A few people — often professional musicians — have the

remarkable gift of absolute pitch. The Reverend Sir Frederick Ouseley (1825-89) observed at the age of five that his father blew his nose in G and, not much later, that the wind was whistling in D. On another occasion he was able to identify a house (having forgotten its number) by the pitch of the door knocker.

Even in normal people the sense of pitch is highly developed. An average person can, over most of the frequency range, detect a change in frequency of three parts in a thousand, corresponding to about 1/20 of a semitone. Between the highest and lowest audible frequencies, about 2,000 levels of pitch can be distinguished. The mechanism controlling this remarkable ability presumably resides in the brain and not in the ear, where perception of pitch appears to be relatively coarse. In the human ear there are about 28,000 nerve fibres connected in a complicated way to the 23,500 hair cells. A moderately loud sound at a frequency of 1,000 Hz will activate about 7,000 nerve fibres. A change in frequency of a semitone will cause 400 of these fibres to become silent while 400 more come into action. We do not know how the brain extracts such precise information from these relatively crude signals.

Indeed, we do not know why man has such a remarkable ability to recognise small changes in pitch. The ability to recognise a wide range of noises did, of course, have substantial survival value for our remote ancestors – but the origin and significance of a sound is more easily recognised from its quality than from its pitch.

The quality of a sound embodies the difference between (for example) a trumpet and a tuning fork. A note of the same pitch sounds quite different when rendered by different musical instruments. This effect occurs because, except for the tuning fork, musical sounds are not pure vibrations of a single frequency but are made up of the fundamental frequency with a mixture of higher frequencies (called harmonics) which combine to give the distinctive tone by which we distinguish one instrument or type of sound from another.

The ultimate microphone

The sensitivity of the ear is remarkable. The movement of the ear drum in response to the minimum audible sound at a frequency of 1,000 Hz is about a thousand millionth of a centimeter — considerably less than the diameter of a hydrogen atom. The movement of the basilar membrane in response to the same sound is about a hundred times smaller — not much more than the diameter of an atomic nucleus.

This level of sensitivity, far beyond the achievement of any microphone, represents the ultimate limit of performance. If the sensitivity of the ear were further increased, we would be conscious of the continual bombardment of the ear drum by air molecules. Even at the present limit, it is surprising that we do not hear the flow of blood through the tiny vessels in and near the ear drum. A rather similar situation has already been mentioned in discussing the eye, where a complicated structure of cells and nerve-endings lying fair and square on the retina is simply not seen. The ear, like the eye, can ignore signals which are constantly present, while retaining its sensitivity to any change. Stories of people who sleep soundly through a steady noise and wake when it stops are quite believable. In noisy surroundings, the ear has the ability to select certain sounds for concentrated attention while ignoring the rest; this is the cocktail party effect, which allows a persistent listener to pick out one voice or one conversation from a babble of talk in a crowded room.

Though the ear can respond to signals not much more than the movement of a single molecule, it also deals quite well with loud sounds. Indeed it is only recently that the advance of civilisation has given us the ability to produce noises loud enough to cause widespread damage to the hearing.

Loudness levels are expressed in decibels (usually abbreviated to dB) above the threshold of hearing. Typical examples are:

rifle fire (at marksman's ear)	160 dB
jet aircraft taking off, noise at edge of airway	140 dB
jet aircraft passing overhead	120 dB
noise in factory	100 dB
alarm clock ringing at 1 metre	80 dB
conversation	60 dB
quiet living-room	40 dB
faintest audible sound	0 dB

The decibel scale is logarithmic – that is, a change of 10 dB corresponds to a tenfold increase in loudness, 20 dB to a hundredfold increase, 30 dB to a thousandfold and so on.

Hazards of noise

As the intensity is increased, the sensation of a loud sound gives way to a painful feeling, which is the warning sign of imminent damage to the delicate organs of hearing. At a frequency of 1,000 Hz the threshold of pain is about 120 dB above the threshold of hearing. In this part of the spectrum the ear can therefore cope with a range of intensity of a billion to one.

Exposure to sounds near the threshold of hearing will, if continued, cause temporary or permanent damage. Up to a sound level of 80 to 90 dB – roughly that of a concrete breaker, a tree saw or a noisy motor cycle at ten yards, range – the noise will probably be annoying but, except for continuous exposure, will not produce any defect or damage. Even at these noise levels, a temporary partial loss of hearing can be detected after a ten minute exposure. As the noise level or the exposure time are increased, the loss of hearing becomes more pronounced and the time required for recovery of normal hearing becomes longer.

The increasing prevalence of pop music, amplified to very high levels (often well over 100 dB in discotheques), has brought the situation where noise, like alcohol and other drugs, can be a source of self-inflicted injury.

The ear does have some built-in protection against excessive noise. The two small muscles in the middle ear, which come into action at noise levels of 80 to 85 dB, stretch the ear drum and increase the resistance to the transmission of sound through the auditory ossicles. Some people can actually contract the appropriate muscles at will, producing a reduction of about 30 dB in sound transmission at low frequencies, though the effect disappears above about 2,000 Hz. The action of the tympanic muscles in suppressing unwanted noise may be likened to that of the iris which restricts the passage of bright light into the eye.

The energy requirements for hearing are quite small. The output of a transistor radio, which is often loud enough to be a nuisance, is well below one watt − about enough to light a torch bulb. The sound energy produced by a million people all talking at once would amount to only 15 or 20 watts − not enough to light a car headlamp.

Sometimes we hear sounds, usually hissing or humming, which are produced entirely inside the head. These noises (known as tinnitus) are almost always experienced during the recovery from exposure to very loud sound and they occur in certain diseases of the ear, but are also quite common in people with normal hearing. Tinnitus may be explained as the discharge of electrical impulses in the auditory nerve, as a response to irritation, but the effect is not fully understood.

Keeping upright

The ears, as well as proving a high fidelity stereophonic hearing system, are also responsible for keeping the body in balance. Our remote ancestors, like today's four-footed creatures, had no difficulty in keeping their balance, whether at rest or in motion. The mechanical design of a four-footed animal involves two arches, one formed by the fore legs and shoulder girdle and the other by the hind legs and pelvic girdle. The two arches are connected by the spinal column and most of the organs and tissues of the body are suspended from the spine or enclosed in the rib cage attached to it. The navigation and early warning systems, including the brain, ears, eyes and nose, are mounted on a universal joint in front of the main structure.

The change to an upright posture brings many advantages, such as the opportunity of developing manual skills, but also brings some obvious difficulties. When lying, sitting or kneeling, the body is stable, but the standing position is, in engineering terms, inherently unstable since the centre of gravity is quite high and the area of contact with the ground (within which a vertical line through the centre of gravity must lie to prevent falling over) is very small in relation to the cross-sectional area of most of the body.

The body keeps in balance with the help of two systems. The first is a complicated network of fluid-filled tubes which the early anatomists, in despair, named the labyrinth. This organ has been compared to the automatic pilot on an air-craft, but the analogy is not very good because the automatic pilot generally relies on external signals, whereas the control system provided by the labyrinth and the rest of the brain is entirely self-contained.

The labyrinth might more reasonably be compared to the inertial guidance system used in missiles and submarines, or, even more simply, to an assembly of three spirit levels. The system does in fact include three semi-circular tubes (usually known as canals) of which one is horizontal, one is vertical in fore and aft direction, and the third is vertical in the left to right plane, so that all three are perpendicular to each other.

In man, the semi-circular canals are lined internally with hair cells and are filled with fluid. They respond to rotational acceleration, as for example when the head is turned, nodded or wagged from side to side. In any of these movements, the canals move along with the rest of the head but the fluid filling, because of its inertia, lags behind and the hair cells are therefore subjected to a drag. Any movement of the head will provoke a response from at least one of the semi-circular canals on each side, but the fluid filling returns to its normal state when a steady speed (or a state of rest) has been reached. The passenger sitting with his eyes closed cannot tell whether his aircraft is moving at constant speed or is at rest on the ground.

Apart from the semi-circular canals, the balance system also contains two small vessels known as the saccule and utricle. These structures act like seismographs, used for detecting earthquakes. The seismograph is a massive pendulum which, because of its inertia, remains at rest during a rapid tremor while the earth around it vibrates. Consequently a pen attached to the pendulum can be used to draw a graph of the earth tremor.

In the utricle and saccule, the hair cells are embedded in a stiff jelly containing crystals of calcite or aragonite (crystalline forms of calcium carbonate or chalk) known as otoliths. Lobsters and other shellfish use grains of sand as otoliths. If lobsters are kept in water containing iron filings, their otoliths can, after a while, be moved by an external magnet – a technique which gives great scope to the ingenious experimenter. In man, however, the otoliths are manufactured from calcium in the diet. The otoliths are denser than the surrounding jelly and tend to stay behind when the head moves. In this way a force is generated on the hair cells and an appropriate message is supplied to the brain.

Although the labyrinth is a very effective system for sensing change of position and therefore for supplying signals to stabilise the position and posture of the body, it is not absolutely indispensable. In Méniere's disease, the operation of the semi-circular canals and their nervous connections is disorganised. Consequently the patient suffers attacks of dizziness and may feel that he is rotating even though he is lying at rest in bed. The disease is sometimes dealt with by cutting the nerves joining the labyrinth to the brain. Patients who have been treated in this way are able to walk about and even to cycle, as long as they keep their eyes open. This situation arises because, in practice, we rely on our eyes to tell us about motion and acceleration rather than on the intricate apparatus of the labyrinth. One example of this ability is a matter of common experience. When one train passes another in moving out of a railway station, a passenger whose view of the platform is blocked by the other train can convince himself, merely by concentrating his thoughts for a second or two, that his own train is stationary – or, equally easily, that his own train is moving.

The mechanisms of the human body serve many different functions but all of them are related to the importance of survival. This aspect of design and performance is obvious in organs and systems which come into close contact with the external environment, for example the eyes, the ears, and the skin. In earlier times, the environment was essentially unfriendly; the progress of man to his present dominant position over other living creatures owes much to the superior abilities resulting from the highly specialised development of the eye and the ear in the course of evolution.

Today, people who live in the developed countries have the power to avoid the discomforts of the environment, but the eye and the ear are still important. Even the most modern communication systems, such as television, computers and satellites, will not work without eyes and ears at each end to prepare the input and interpret the output.

The ear, though designed for a very different environment, contributes uniquely to the life of modern man. In the course of time, the ability to distinguish the noises of the forest conferred a better chance of survival; today this ability has developed to allow sophisticated forms of musical expression and all the subtleties of the spoken word.

The ears, in association with the brain, provide us with high fidelity stereophonic sound with a range extending from a whisper or the rustle of a leaf to the loudest natural noises; it is only in recent times that the electronics and aircraft industries have achieved exposure of large numbers of people to sounds beyond the normal limits of loudness – and, even there, we find that the ear has a defence mechanism.

The success of the ear as the organ of balance is seldom fully appreciated. The upright posture which has allowed man to develop technology and artistic expression far beyond the achievement of any other animal would not be possible without the elaborate and subtle examples of feedback and automation which allow us to maintain our balancing ability. The combination, in such a small space, of the hearing and balancing systems of the body represents a remarkable achievement of biological engineering.

7: perfumed air

'As plain as the nose on your face'

The nose may be conspicuous but, as to design and operation, it is anything but plain. The space behind the nostrils contains the world's finest air conditioning plant, combined with a detection system of extraordinary sensitivity, which analytical chemists are not yet able to explain, still less to imitate.

The nose, like the ear, has two quite different functions. Firstly, it cleans, warms and humidifies the air that we breathe, not only to protect the delicate lining of the lungs from thermal or chemical insult, but also to ensure that the air that reaches the alveoli, down in the lungs, is in the best condition for the exchanges of oxygen and carbon dioxide which are essential to life. Secondly, the nose contains a patch of yellow or brown tissue, no more than a square inch in area, where the sense of smell is concentrated.

The human nose, though greatly surpassed in sensitivity by many animals and insects, is still a remarkable detector. Ethyl mercaptan (a foul smelling substance used by skunks) can be detected at a concentration so low that a drop no bigger than a pinhead would fill the Houston Astrodome's forty million cubic feet.

Taste and smell are closely related (both senses are impaired by a cold in the head) but taste is not as sensitive as smell. Strychnine hydrochloride, a bitter chemical, can be tasted more readily than almost any substance known, but the minimum amount that will affect the taste buds is about one hundredth of a microgram or ten thousand times more than the threshold for detecting ethyl mercaptan by smell.

Breathing

We need air conditioning in the nose because the air that we breathe is usually too cool and too dry to do its job efficiently in the lungs. The nose will cope with air at temperatures down to −40°C and, if need be, up to +40°C or more. The humidity problem is a little more complicated. The atmosphere always contains a certain amount of moisture. At any particular temperature, the amount of water vapour in the air can be increased up to the saturation point, at which condensation begins. The air is then said to have a relative humidity of 100%; in other conditions the relative humidity is the amount of water vapour present in the atmosphere expressed as a percentage of the amount that would cause saturation at the same temperature. On a wet day, the outside air is saturated; on a dry day the relative humidity may be about 50%.

Air for delivery to the lungs must be warmed to the body temperature of 37°C and must be saturated with water vapour for efficient exchange of carbon dioxide with the circulating blood. As the temperature increases, air needs more water vapour to saturate it; consequently air taken in at normal room temperature has to be provided with both heat and moisture before it reaches the lungs. Once past the nostrils, the inspired air takes a rather tortuous course among the turbinal bodies – three corrugated structures, one of bone and two of cartilage. The rough surfaces of these structures break up the air stream and result in turbulence which stirs the air, bringing it into effective contact with the mucous lining of the respiratory tract and allowing rapid warming. This process goes on until temperature equilibrium is reached, with the inspired air and the nasal mucosa both at a temperature of perhaps 30-32°C; further warming and humidification occur between the nose and the lungs.

On its return journey, air leaves the lungs at a temperature of 37°C and relative humidity of 100%. On reaching the nasal mucosa, previously cooled by the inspired air, it gives up some heat. When once its temperature drops below 37°C, the expired air is more than saturated and accordingly sheds some water as condensed droplets. Some of the energy used to evaporate the water needed to moisten the inspired air is now returned as latent heat with the water condensed from the expired stream.

The cooling process is not complete and the air usually leaves the nose at a temperature of about 32°C, or a few degrees less in very cold weather. About 25% of the heat and water taken up by the inspired air is returned with outward breath; the remainder is lost as the expired air mixes with the rest of the atmosphere. An adult man in a temperate climate loses about 250 ml of water and 350 kcal of heat in the expired air every day.

A patient deprived of the use of his nose can manage without a sense of smell, but the loss of the air-conditioning ability leads to trouble after only a day or two. In cases of polio, where the muscles used in breathing are paralysed, air from a mechanical respirator of the iron lung type is delivered, under pressure, straight into the trachea (windpipe) from where it passes to the patient's lungs. The warming and moistening performed so effortlessly by a normal person are difficult to achieve in machine-assisted breathing. Consequently the inside of the trachea, near the entry of the air-tube, becomes dry and caked, requiring regular attention to avoid discomfort.

Attempts to make an artificial nose have not been altogether successful. One method uses a roll of fine-mesh silver gauze, in a tube of plastic material, as the medium for exchange of heat and water vapour; this technique works moderately well. Heat is readily taken up and released by the silver, and the fine mesh provides a convenient base for the alternate deposition and evaporation of water. However, the artificial nose must be kept at a temperature close to 37°C, requiring a water-jacket, heater and thermostat. These arrangements are difficult to provide since the artificial nose loses its purpose if it is not situated very close to the air inlet to the trachea.

The difficulties met in devising an artificial system of this kind remind us that the air conditioning system of the nose is, in engineering terms, very well designed. The working substances, air and water, are ideal for the purposes that they serve. The specific heat of water (that is, the amount of heat required to raise the temperature by 1°C) is one calorie per gram, a very high value exceeded by no liquid except ammonia. Because of the high specific heat, a small amount of water can store quite a lot of heat; the moist lining of the respiratory tract therefore makes an effective reservoir of heat for the incoming and outgoing air. The specific heat of air is about 0.24 calorie per gram, quite a low value. For this reason the temperature of air can be changed substantially by the addition or subtraction of only a small amount of heat. The warming of the incoming air can therefore be effected without seriously lowering the temperature of the nasal mucosa.

Though the system is designed mainly for cool or temperate climates, it works quite well at high temperatures. If the incoming air is at a temperature above 37°C, and is not saturated with moisture, it will be humidified by the evaporation of water from the mucosa. The latent heat (that is, the amount of energy needed to evaporate the water) is very high (about 0.58 kcal per gram) and the loss of heat on this account is more than enough to outweigh the warming effect of the hot incoming air.

At really low temperatures, the performance of the nasal air conditioner is remarkable. During the Second World War, a fighter pilot lost the canopy of his aircraft and flew for two hours at a temperature of −30°C against an air stream travelling at 250 mph. Some of the tissues of his face were destroyed by frostbite − but the only damage to his respiratory tract was a sore throat.

'Hellish dark' said James Pigg, 'and smells of cheese'.

The huntsman (in the famous novel *Handley Cross* by R. S. Surtees) had been discussing prospects for the next day's sport. His comments on the state of the weather were somewhat inaccurate because, confused by the darkness and fumes of brandy, he mistook a cupboard for a window — demonstrating that smell is sometimes a keener sense than sight.

Smell was probably the first of the remote senses to appear on the evolutionary scene. Long before there was any need for good sight or hearing our remote ancestors, swimming in the sea, developed a sense of smell to help them in finding food and avoiding predators. It might be thought that a fish, which does not breathe air, uses taste rather than smell. However, almost every fish has an organ anatomically similar to the nose, as well as taste buds all over its skin. Even in man, smell is essentially an underwater sense, since it functions in the film of moisture lining the appropriate parts of the nasal passage.

The sense of smell is still highly developed in fish and is probably responsible for the accuracy with which salmon return from the open sea to the rivers in which they were born. Many fish hunt at night and have only smell to guide them to their prey; others, including the dogfish (which appears in tins as rock salmon), hunt entirely by smell even through the day. In man, the sense of smell begins in two small patches of coloured mucosa each about half a square inch in area, located (one on each side) at the top of the nasal passage.

The olfactory mucosa contains (on each side of the nose) about 50 million receptor cells, in which originate the signals eventually interpreted by the brain as the sensation of smell. The receptor cells are relatively long and thin, but each is surrounded and supported by larger cells which seem to be there only as packing material.

Each receptor cell ends in a bundle of fine hairs, known to the anatomist as cilia, which have the power of spontaneous movement. They float in or on the film of mucus which lines the nose. Signals from the receptor cells undergo some initial data processing before being transmitted, along the 50,000 fibres of the olfactory nerve, to the brain.

There is nothing in the process of smell corresponding to the lens of the eye, or to the ear drum, middle ear and cochlea which separate the brain from the outside world with regard to hearing. Smell signals pass directly to the brain, an arrangement which reminds us of the great importance of smelling power at earlier stages in man's evolution.

For the sensation of smell to be evoked the substance concerned must be drawn over the olfactory mucosa and a little of it must stick. In the normal way, most of the air that we breathe does not pass through the olfactory region — but, in sniffing, a slight rearrangement of the muscles allows us to deliver a suitable sample (about 20 ml) to the right place. Alternatively, vapours or fragments of material can reach the olfactory region from the back of the throat; the sense of smell, excited in this way, supplements taste in our appreciation of food.

Smell, like vision and hearing, responds quickly to a change of stimulus but adapts after quite a short time to a constant stimulus. This is why a person walking into a smoke-filled room or railway carriage will complain, although people already there do not notice the atmospheric pollution.

Theories of smell

During the last century there have been more than thirty major attempts to provide a scientific explanation for the mysteries of olfaction. All of the theories which have been offered use the same starting point: the interaction between molecules of the material to be detected and the cells in the small region of the nose where the sense of smell resides.

Some theories depend on the plausible suggestion that chemical structure is the key. One long series of experiments carried out by G. M. Dyson, a chemist interested in perfume, began by studying phenyl mustard oil, a pungent synthetic compound, chemically related to a material extracted from mustard seeds. Dyson made a whole series of chemical compounds, all slightly different, by adding chlorine, bromine, iodine, and other atoms at various places in the original molecule. The smell of each new substance was recorded in the hope that some simple relation might be found with the chemical structure. Chemical properties of the various compounds were also studied in the hope of finding some correlation with smell. There was some success in this investigation but on the whole it appeared that the chemical structures of properties were not closely related to smell.

A curious theory proposed in 1947, and soon discarded, suggested that the olfactory nerve cells in the nose gave off infra-red radiation; if the appropriate chemical substance happened to be passing at the time, this radiation would be strongly absorbed, thereby cooling the nerve endings and giving rise to the sensation of smell. A serious objection to this proposal is that a transfer of heat between the nerve cells and the surrounding air can only take place if there is a difference of temperature. In practice, however, it is found that the sense of smell is still intact even when air at a temperature of 37°C is breathed. It is difficult also to see how very tiny changes in temperature, corresponding to extremely low concentrations at which many pungent smells can be detected, could be distinguished from the random changes in temperature produced even by the passage of pure air.

Another proposal, made more recently, depends on a classification of smells. This is not as easy as the classification of colours, sounds or even tastes, but a number of attempts have been made. In one scheme four kinds of smell are postulated − fragrant, acid, burnt and caprylic; the last group (named from the Latin word for a goat) includes most of the foul and disagreeable odours.

Any particular smell may contain more than one of these groups in different proportions. Perfume manufactures often use a scale of 0-8 on which, for example, the rose is represented by 6423, meaning six parts of the fragrant component, four of the acid component and so on. It may seem surprising that the smell of a rose contains a proportion of foulness, but many substances which have loathsome odours in concentrated form are not disagreeable when greatly diluted. In the blending of perfumes it is quite common to use small quantities of musk (an animal secretion), civet (a similar material) or ambergris (undigested remnants from the intestines of a whale). These materials, though not themselves pleasant-smelling, are considered to improve the perfumes to which they are added.

Since perfumes often contain a small proportion of offensive odours, it might be expected that foul smells include traces of fragrance. The question has not been studied very seriously − since there is no commercial incentive to imitate an unpleasant smell.

Many classification schemes have been produced (usually with more than four basic odours) and some of them are found quite serviceable in scientific or industrial work. The use of these empirical schemes and classifications naturally leads to the speculation that there may be specialised receptor cells in the olfactory mucosa, some sensitive to one basic odour and some to another. There is of course, a parallel in the ear, where different regions of the basilar membrane (page 82) respond to notes of different pitch.

J. E. Amoore suggested in 1952 that seven main kinds of smell are associated with vacant sites of characteristic shape on the surface of the olfactory mucosa. Most of the sites are envisaged as circular or elliptical depressions but some are more complicated. The appropriate smell will be evoked when a molecule of the right shape drops into one of the vacant places. A hemispherical depression is associated with a camphoraceous smell and a wedge shaped cavity with a minty smell. These ideas recall the speculations of Lucretius, first elaborated more than 2,000 years ago:

> And note, besides, that liquor of honey or milk
> Yields in the mouth agreeable taste to tongue,
> Whilst nauseous wormwood, pungent centaury,
> With their foul flavour set the lips awry;
> Thus simple 'tis to see that whatsoever
> Can touch the senses pleasingly are made
> Of smooth and rounded elements, whilst those
> Which seem the bitter and the sharp, are held
> Entwined by elements more crook'd, and so
> Are wont to tear their ways into our senses,
> And rend our body as they enter in.

Amoore's classification received some support by the finding that, for a number of synthetically prepared chemicals, smell is correlated with the shape of the molecule; there were however a great many difficulties and much of the theory was abandoned in 1965. The relation of molecular size and shape to the odour of the corresponding material is however generally accepted, even though no exact numerical or geometrical classification has yet been found satisfactory.

Another tantalizing theory, proposed in 1954 by R. H. Wright, stems from the observation that the region in the nose associated with the sense of smell has a distinctive brown or yellow colour, due to the presence of an olfactory pigment. This pigment is found in relatively large amounts in those animals which have a highly developed sense of smell. In albinos (who have no skin pigmentation) the sense of smell is usually deficient or absent.

Colour in any object is associated with the presence of electrons which can absorb energy rather easily from an incident beam of light, thereby passing to higher energy states. The coloured patch inside the nose is not likely to be excited by light but a similar effect might be produced by absorption of energy from chemical reactions or from neighbouring molecules.

Wright suggested that an electron in a molecule of olfactory pigment might be encouraged to jump (and in doing so to generate a signal which could be passed along the olfactory nerve to the brain) by the proximity of another molecule which had the appropriate natural frequency of vibration. The effect evoked here is resonance, as in the legendary tale of the singer who could shatter a wine glass; it will be recognised that the theory has some affinity with the resonance theory of hearing.

Like most theories of smell, this one fits quite well with some of the experimental evidence and fails to account for other observations. As a result of a great many calculations and experiments, Wright suggested that the olfactory pigment should have a chemical structure similar to that of vitamin A. It has, curiously enough, been known for some time that rats deprived of vitamin A appear to lose their sense of smell. Two scientists in New Zealand found that the olfactory regions of the cow contained vitamin A and compounds chemically related to it. When large doses of vitamin A were given to fifty-six patients who had lost their sense of smell, fifty of them had a partial or complete recovery. These clinical findings do not, however, appear to have been confirmed in other places.

Since the primary physical or chemical changes leading to the sensation of smell occur in the moist surface of the nasal mucosa, it might be supposed that the application of a ready-made solution would be even more effective – particularly since it is known that a fish smells in this way. It is not easy to ensure that the appropriate region in the nose is covered by a solution, but reliable techniques have been developed and a test has been made on many occasions, with odorous substances in water or in salt solutions closer in composition to normal biological fluids. The surprising conclusion from this work is that the presentation of the stimulus in the form of a solution does not enhance the sense of smell but reduces, distorts or obliterates it. It is interesting also to note that aquatic mammals such as the dolphin and the porpoise, which originally developed on land but in the course of evolutionary history have returned to the water, have little or no sense of smell.

Smell and the engineer

As a technical problem, the sense of smell involves two branches of engineering. The chemical engineer finds much to admire and cannot give a comprehensive explanation, but he recognises one or two landmarks. It seems fairly clear that molecules of an odorous substance are adsorbed by the sensitive region in the nose. Adsorption is a surface effect, in contrast to absorption, which involves the spread of one substance in depth through another; water is absorbed by a sponge, but adsorbed by the paper on the bathroom walls. The molecules of material which are relatively far apart when milling about in three dimensions become much closer together when adsorbed onto a surface. This concentration effect gives a plausible explanation for the ability to detect some materials at extraordinary low concentrations in air.

The adsorption of a gas or vapour onto a solid surface is a process which invariably releases heat, so that a supply of energy is available to stimulate the olfactory nerve; there could of course be no transmission of messages to the brain without some expenditure of energy.

The electronic engineer observes that the nose contains a great number of receptor cells, in which the process of smell is begun. There are about fifty million of these cells in the two small areas of olfactory mucosa (one on each side of the nose) with complicated connections to about 50,000 nerve fibres communicating with the region of the brain where the sense of smell is localised. The great number of receptor cells probably accounts for the excellent sensitivity and allows for the possibility of sophisticated data-processing of incoming signals (comparable with what goes on in the eye and the ear) to give good discrimination. As we have already seen, the sense of smell operates without any obvious processing organ (such as the retina of the eye or the cochlea of the ear) between the incoming signal and the brain.

How many separate smells can we recognise? The problem of classification has already been mentioned. The four-component system (fragrant, acid, burnt, caprylic), with the rose at 6423 and other odours represented by numbers between 0000 and 8888, is quite useful. A more logical scheme, with eight primary odours, was proposed as long ago as 1895 by Alexander Bain, Professor of Logic at Aberdeen University. This classification is worth quoting.

1. Fresh odours, those that have an action akin to pure air, or coolness in the midst of excessive heat . . . many of the balmy odours of the field and garden have this effect.
2. Close or suffocating odours — the opposite of freshness.
3. Sweet or fragrant odours, representing the pure or proper pleasures of smell — the odour of violet is a pure instance.
4. Malodours. The opposite of sweet; the brown scum of a stagnant pond.
5. Pungent odours, such as ammonia or nicotine.
6. Ethereal odours, such as alcohol and chloroform.
7. Appetising odours; the smell of flesh excites the carnivorous appetite and rouses the animal to pursuit.

Other classifications include larger numbers of odours, but none of the schemes succeeds in identifying every kind of smell. Ernest Crocker, the American chemist who devised the four-number system, could himself recognise 9,000 separate smells — but he had exceptional powers and was, indeed, employed for many years as a smell consultant. The average person can recognise between 2,000 and 4,000 different smells; this figure may be compared with the discriminating power of the eye (about 200 colours) and of the ear — about 2,000 levels of pitch.

In man, however, the importance of smell has declined during evolutionary progress. The primitive brain was largely concerned with smell; seeing and hearing were later developments. Many animals still rely on smell as their main contact with the rest of the world. The dog, for example, has poor eyesight and quite good hearing – but most of his brain is used in connection with smell.

The area of the olfactory mucosa is about fifty times as large in a dog as in a man. The dog's nose is not much bigger than our own, but the sensitive mucosa is greatly ridged and folded, allowing a considerable surface area to be fitted into a modest space.

The limit of concentration at which various substances can be detected is always lower for a dog than for a man. For some chemicals a dog is about a thousand times more sensitive than a man, but for many other substances the dog's threshold of perception is a million times lower. It might be thought that an animal with such a sensitive nose would be overwhelmed by smells of normal intensity. This does not happen, for the brain has effective methods of reducing the sensitivity when a strong stimulus appears; a person with unusually acute hearing is not deafened by the sounds of everyday life.

Where the dog has a thousandfold advantage he is probably smelling the same smell and using rather similar equipment to do it. The dog's nose is designed to give freer access of air and his brain is probably better able from experience to deal with the signals that reach it. When the sensitivity is a million times better, a dog is probably using powers that are not found at all in the human subject.

Foxhounds, police dogs and, more recently, the dogs trained to sniff out marijuana and other narcotics in piles of luggage, all display smelling powers that the human nose cannot imitate. A dog can recognise a person by his smell – indeed he can do a great deal better, for, having smelt someone's hands, he can afterwards recognise the same individual by his underarm odour. Probably we all exude a characteristic mixture of odours which, though seldom strong enough for other people to recognise, provides a chemical fingerprint readily perceived, classified and remembered by a dog.

When two dogs meet, the process of recognition and

greeting is by smell and not, as with humans, by touch, sight and sound. In regions not yet domesticated, dogs and other animals mark the boundaries of their territory by urinating. In view of the dog's undoubted ability to recognise people by smell, this simple method of identification and warning is quite effective. Even in built-up areas a dog uses his nose to tell which of his fellow-creatures have passed the same way.

The sense of smell in insects is even more remarkable. The female silk-moth will attract males from great distances. In one experiment, a female was positioned under a gauze hood in a field and a number of males, suitably marked, were taken by train to a distance of 2½ miles away. 40% of them made their way back to the female when released; a return of 26% was achieved over a distance of seven miles. It is difficult to explain these experiments, since the concentration of the attractant substance released by the female moth can hardly have been more than one molecule every few centimetres at a distance of a few miles. There is, however, little doubt that the sense of smell is the critical factor here. If the female moth is covered by a glass jar, the males are quite incapable of finding her. On the other hand a piece of paper which has touched the scent gland on the female moth will act as an effective target.

Taste

Smell, like hearing and vision, is a long distance sense which tells us about the environment and gives early warning of danger. Taste is different, for it is a very short range sense operating only on food already in the mouth. The energy available for stimulating receptor cells and generating a message to the brain is very much greater from a bite of food than from a distant sound or sight or a whiff of some distinctive odour. It is therefore not surprising that taste is much less sensitive than smell.

For ethyl alcohol, one of the few substances which has a taste independent of its smell, the threshold for tasting is about seven hundred micrograms and for smelling about twelve micrograms. A solution of alcohol in water has no taste unless the concentration is at least 14%; this is more than the level found in beer and unfortified wines, but the enjoyment of these beverages does not depend on the taste of alcohol.

Though taste and smell are closely related senses, they can be studied separately by pinching the nostrils, so that air – even from the back of the throat – cannot pass over the olfactory mucosa. In this state, most people cannot distinguish ham from lamb, apple from pear or claret from vinegar.

In the common cold the sensitive hairs protruding from the receptor cells in the nasal mucosa are liable to be put out of action because the normally watery mucus becomes much thicker. Strictly speaking, the sense of taste is not affected, but in practice the taste of food is greatly impaired; this is because much of the pleasure associated with food and drink is attributable to vapours which rise from the back of the mouth into the nose and are smelt rather than tasted.

Scientific tests demonstrate that there are four basic tastes – salty, sour, sweet and bitter. Recognition of these four tastes is to some extent concentrated at particular locations around the edge of the tongue. The taste buds responsive to sweet substances are mostly near the tip, while bitter tastes are detected at the back of the tongue. Salty flavours activate taste buds at the sides of the tongue, near the tip, and sour tastes are appreciated on each side of the tongue, behind the salt-sensitive areas.

Threshold levels for the four basic tastes differ somewhat among individuals but are approximately as follows:

Sour (hydrochloric acid)	–	0.007%
Salt (common salt)	–	0.06%
Sweet (sugar)	–	0.7%
Sweet (saccharin)	–	0.001%
Bitter (brucine, an alkaloid related to strychnine)	–	0.00005%
Bitter (strychnine)	–	0.0002%

The taste buds are quite numerous, about 9,000 in man, almost all on the tongue, though a few are found on the soft palate and in the throat. In children, the central part of the tongue and the insides of the cheeks are also well supplied with taste buds but these disappear in adolescence. Children have a well-developed sense of taste, particularly for sweet things. The sense of smell is, however, not so fully developed in children; it is for this reason that they usually do not favour highly seasoned or aromatic foods.

Taste, like other senses, deteriorates with advancing age. The taste buds are fully developed in adolescence and remain fully effective up to the age of forty-five. After that they begin to atrophy and the sense of taste becomes less acute. There are no differences in the sense of taste between man and woman. A baby has more taste buds than an adult, but they are not fully developed functionally. Thanks to the greater permissiveness that now characterises infant feeding, it has been learnt that babies can recognise and enjoy fruity or meaty tastes from an early age.

The number of taste buds differs greatly between one species and another. The cow has 35,000, which seems a lot for an animal living on such a monotonous diet as grass. The larger tongue might be offered in explanation – but the rabbit has 17,000 taste buds, while the pig and the goat have 15,000. The hare, though similar to the rabbit in many ways, has only 9,000, the same as man. Birds have few taste buds – usually less than 100.

Experiments with dogs, rats and other animals show taste thresholds roughly the same as in man, with a few interesting differences. Cats do not taste sweet things and saccharin tastes bitter to a dog. Rats will avoid certain poisons, even at concentrations too low to harm them – but will sometimes drink lethal doses if the poison is sweetened with sugar.

Why do some animals have more taste buds than others? The question does not yet have an answer. The number of taste buds does not seem to influence the sensitivity to taste, or the ability to distinguish between wholesome substances and poisons.

The sensory organs of taste are similar to those responsible for smell. Each taste bud contains a number of receptor cells and each cell ends in a tiny hair which projects into the mouth. At the opposite end, each cell is joined to a nerve fibre leading to the brain.

Unlike every other part of the nervous system, the taste receptor cells have a very short life, measured only in days. A taste cell develops deep in the skin and then moves into the taste bud where it first connects with a nerve ending. In view of the short life of the taste cells, it is puzzling to find that many of them appear to have specialised tasks, some responding to salty tastes and some to sweet, sour or bitter substances. The most likely answer is that the taste cells are all made to a general-purpose specification and that the specialised response is programmed by instructions delivered from the nerve; in this connection it is significant that, if a nerve fibre is cut, the corresponding taste cell dies.

Smell and taste are examples of sophisticated chemical technology, based on rapid and sensitive analytical methods. The analytical chemist usually takes his samples to the laboratory for investigation. Sometimes it is important, especially in the control of industrial processes, to do the analytical work almost instantly, so that any necessary corrections can be applied to the production line.

On-line analysis (as this method of working is known) is used where speed, and ability to operate on the factory floor, are more important than high sensitivity. Smell and taste are examples of on-line analysis, often achieving extreme sensitivity. A chemist, working at leisure in his laboratory, could detect some substances (for example, alcohol and sugar) at levels below the threshold for taste and smell, but he could not equal the sensitivity of the nose in detecting ethyl mercaptan. Still less could he approach the performance of dogs or insects.

When a chemical analysis is made, the results have to be interpreted. An aspect of taste and smell which defies emulation by the chemist is the great rapidity with which information can be processed and judgements made; a simple taste (bitter, sour, sweet or salty) can be recognised in only a four hundredth of a second. The instant analysis that we all perform effortlessly with the tongue or the nose (and the associated regions of the brain) and the speed with which we can detect different smells in succession are achievements that cannot yet be equalled in the laboratory or the chemical process plant.

When dog meets dog, the message that passes is: smell me. But when a man meets another man he says: tell me. When he has listened he confirms his understanding by saying: I see. Over the course of evolutionary development, man has come to rely on his eyes and ears as the main links with the outside world and with other people. Man's superior place in the living world owes much to the unsurpassed combination of skills that he has acquired in hearing and in seeing.

But we have not altogether abandoned our inheritance from the days when smell was virtually the only sense and the brain was little more than an extension of the nose. Today we have access to a variety of food and drink, more extensive than our remote ancestors ever knew. The primitive senses have adapted successfully to the challenge. Taste and smell, no longer a matter of life and death in unfriendly surroundings, contribute greatly to our health, happiness and enjoyment of the world around us.

8: the vital stream

The blood is one of the most interesting systems in the body. As a tissue, it has a distinctive structure, many specialised chemical reactions and a long list of essential tasks. In comparison with all the other tissues of the body it has two unique properties: firstly it is a liquid and secondly it is always on the move.

The body of an average man contains about six litres of blood. The heart, when resting quietly, pushes out about 100 ml at each stroke, so that the blood makes a complete circuit of the body roughly once a minute. In its journey the blood shows us many fascinating examples of fluid mechanics, chemical engineering and environmental control. To trace the development of this ingenious system, and to see why the body needs blood at all, we need to go back to the primitive ocean which was the home of our earliest ancestors.

Every form of life (save for some specialised micro-organisms) depends on two things — food and oxygen. The oxygen is essential for the chemical breakdown of the food, in order to provide energy or to give the raw materials needed for growth and repair. A waste disposal system is also needed to deal with the unusable residues after the oxygen has done its job.

The simplest animals are single-celled creatures living in the sea, which provides an endless supply of food and carries away the garbage. The inward movement of oxygen and the outward passage of carbon dioxide (the principal waste product) are maintained by diffusion, a simple process depending on the difference in concentration between inside and outside. Sea water is richer in oxygen, so oxygen moves into the cell; the concentration of carbon dioxide inside the animal is greater than in the surrounding water and carbon dioxide therefore moves out.

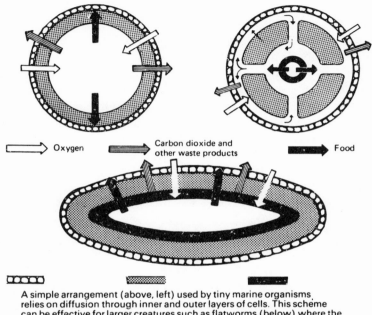

⟹ Oxygen   ⟹ Carbon dioxide and other waste products   ⟹ Food

A simple arrangement (above, left) used by tiny marine organisms relies on diffusion through inner and outer layers of cells. This scheme can be effective for larger creatures such as flatworms (below) where the total thickness of tissue is still limited. Otherwise, increase in size requires the development of an internal circulation system (above, right).

Figure 5 distribution of food and oxygen and disposal of waste products

This is quite a satisfactory arrangement for small animals, but has the limitation that diffusion is not really effective over distances of more than about one millimetre. Suitably shaped animals, including flatworms and a good many marine organisms, have achieved a modest degree of complexity in spite of this restraint, but all higher forms of life require some kind of circulation.

The human animal relies on circulating blood to do several jobs, of which the most important is the collection of oxygen from the lungs and its distribution to the other tissues of the body. While this is going on, carbon dioxide is removed and returned to the lungs where it is lost in the expired air. One part of the circulation takes food molecules, absorbed from the digestive tract, to the liver – the main processing plant for the breakdown of food and its subsequent remodelling into all the complex materials needed to keep the body going. The waste products of this remodelling are collected by the blood and delivered to the kidneys where, after further processing, they are excreted in the urine. The blood acts as a transport system in other ways, taking fat from the body's stores to the liver as an additional source of energy when food is short, and conveying hormones, or chemical messengers, from the organs where they are made to the tissues where they do their work.

Some of the cells in the circulating blood provide the body's main defence mechanism against invading germs. Other constituents of the blood provide an ever ready repair system. Finally, the circulating blood helps to adjust the body's heat losses, conserving energy on a cold day and shedding it in hot weather so as to keep the temperature of the internal organs within narrow limits.

Since the blood is both a body tissue and a transport system, analysis of a sample will always reveal materials which are merely taking a ride from one part of the body to another. However these are easily distinguished from the essential constituents.

If a sample of blood is left standing in a test tube, it will before long separate into two fractions – a dark red clotted mass, mainly composed of cells, and a clear straw-coloured liquid (known as serum) containing a number of mineral salts and proteins.

The cells are of three main kinds. The red cell is, strictly speaking, not complete in the biological sense since it contains no nucleus. For this reason it cannot reproduce itself by dividing; when it is worn out it simply breaks up and is swept away, to be replaced by a freshly made cell.

Red cells are manufactured in the bone marrow. In early

life most of the bones in the body take part in this process; in the adult, however, red cell production is concentrated mainly in the skull, spine, breastbone, ribs and pelvis.

The red cell passes through several stages of development before reaching maturity. During this sequence of changes it loses its nucleus and takes on its characteristic shape – a disc, somewhat thicker at the rim than at the centre.

Red cells have an average life of about 120 days and are quite numerous, more than five million per cubic millimetre in the average man and slightly less in a woman.

Oxygen transport

About one third of the mass of the red cell consists of haemoglobin, a protein with a remarkable ability to combine with oxygen. This property allows the blood to perform one of its most important functions – the carriage of oxygen. Our utter dependence on oxygen is responsible for many features of the body's engineering design. A man can live for a day or two without water and for several days without food – but for only a few minutes without oxygen. If the brain (but not the rest of the body) is deprived of oxygen for a few seconds, the result is a faint; if the deprivation continues for more than a few minutes, permanent damage is done.

Most of the materials that the body needs to sustain life (food and water, for example) can be stored, but oxygen has to be delivered continually. It might be thought that the evolutionary progress of higher forms of life would have included the development of a chemical (or a specialised tissue) providing oxygen in concentrated form, thereby easing man's dependence on regularity of supply. But there is no such chemical and no such tissue; oxygen has to be transported to every cell of the body in a continuous stream.

A liquid delivery network is the best way to meet this exacting requirement, but the problem is more difficult than it appears. The chemical engineer's first solution would be to use oxygen dissolved in the circulating blood. Unfortunately the blood plasma (that is, the fluid component without the cells) does not dissolve oxygen very readily and holds less than 2% of the body's needs. The gap cannot be bridged by circulating the blood at a greater rate, because that would need larger veins and arteries, a larger heart and a great deal more effort.

Haemoglobin has a great affinity for oxygen and the modest amount present in the blood (about a kilogram) carries enough oxygen to keep all the vital processes going. Having discovered haemoglobin, the engineer might first think of dissolving it in the blood plasma as an efficient way of distributing oxygen throughout the body. This approach would not be satisfactory; a kilogram of protein dissolved in about three litres of plasma would produce blood with the consistency of thick soup — too thick for the heart to pump around the body. Some animals (such as the earthworm) do have haemoglobin dissolved in the circulating blood, but only in relatively small amounts.

In man and many other animals the dilemma is resolved in a very ingenious way. Instead of being dissolved in the blood, the haemoglobin is concentrated in the red cells, which are small enough to be swept along without much additional effort. Each one is covered in a membrane which allows the free movement of oxygen in both directions, but keeps the protein from direct contact with the plasma. In this way the oxygen-carrying capacity of the blood is increased to about sixty times as much as would be provided by the plasma alone.

The red cells are not cells in the usual sense of the word because they have no nuclei and no genetic material. Strictly speaking, they are not living matter at all, but specialised chemical processing plants.

The body's defenders

The white cells are larger than the red cells but not so

numerous; each cubic millimetre of blood contains about 7,000. They have nuclei and have the power of independent movement. The white cell moves like the amoeba, usually the first creature studied in the biology class. Some of the cytoplasm (the jelly surrounding the nucleus) flows in a new direction and the rest of the cell follows it. In this way the white cells move freely in the blood vessels and even through the walls of the small vessels into the tissues of the body.

White cells are of two main kinds, with different sources and different tasks. The granulocytes (so named because their cytoplasm is dotted with dark granules) are made, like the red cells, in the bone marrow. Their main purpose is to help in the defence against infection.

Infection is sometimes thought of as a serious emergency resulting from contact with dirt or disease or from inexplicable misfortune. The fact is that our bodies are permanently infected with a great many microbes, including colonies of those responsible for all of the common diseases. A few places, including the bladder and the lungs, are sterile in the healthy person and the exposed surface of the eye is remarkably clean; the steady flow of tear fluid (which contains antiseptic substances) and the frequency of blinking (about every three seconds) combine to wash germs, dead or alive, into the nose.

Though the skin harbours a thriving population of germs, it keeps them in their place and does not allow penetration into the body. Occasionally, however, foreign germs do gain a hold – perhaps through a scratch or cut or through contaminated food.

In this situation the body's first line of defence is provided by white cells in the circulating blood. Whenever the body is invaded by foreign protein (such as bacteria), granulocytes are mobilised and rush to the scene. Putting out feelers of cytoplasm they surround, engulf and destroy the invading organisms.

The pus which appears at the site of an infection (for example before a boil bursts) is a collection of granulocytes, dead bacteria and remnants of damaged tissue.

Granulocytes also appear wherever tissue is damaged, after a cut, a bone fracture or any other injury. They break up the dead tissue and remove it from the scene, so that new growth can make good the loss.

The other main group of white cells are the lymphocytes, which represent about a quarter of the total white cell population. Some are made in the spleen and others in the lymph nodes.

Leaky pipes

Lymph is one of the body's fluids and is a pale shadow of blood, basically a liquid resembling blood plasma. Its presence reminds us of a surprising design weakness in the blood circulation. It has already been mentioned that white cells can slip through the walls of small blood vessels and wander among the tissues of the body. If white cells can escape, it is to be expected that water molecules (which are much smaller) and molecules of other substances will also be able to leave the circulating blood. This is exactly what happens; indeed some of the constituents of blood are actually squeezed out of the capillaries, the small vessels where the fluid pressure driving the blood is relatively high.

This liquid forms the interstitial fluid and amounts to about eight litres in a normal adult. The veins and arteries of the blood circulation are, in engineering terms, full of leaks. Why do they not dry up, if they are continually losing liquid?

The engineer faced with a leaking circuit might first try to plug the leaks — but if that failed he would look for ways of collecting the lost liquid and returning it to the circulation. The interstitial fluid is dealt with in the same way. Some of it, squeezed out of the circulation at points of high pressure, finds its way in again at places where the circulating blood is under low pressure. Some passes into thin-walled tubes resembling blood vessels but actually belonging to a separate network. The interstitial fluid which passes through is called lymph.

The lymphatic system is a complex network of vessels, usually lying close to veins or arteries. The system has no pump corresponding to the heart, and the lymph moves rather slowly, propelled by the movement of muscles and guided by little flaps of tissue acting as valves. Eventually the lymph goes back into the bloodstream by way of the subclavian veins behind the clavicles or collarbones.

Antibodies and immunity

Lymph nodes are small lumps of tissue located at many places in the body. Their main task is the production of lymphocytes, which then circulate in the lymph, eventually reaching the blood. These cells are also made in the spleen, from which they pass directly into the blood. As we have seen, the granulocytes attack invading organisms (for example in infectious diseases) by swallowing them. The lymph nodes provide a further kind of defence by producing antibodies. Whenever foreign protein enters the body (most usually as the organisms of some infectious disease, but sometimes as transplanted tissue), an immune reaction is provoked. The invader is recognised and antibodies are manufactured in the lymph nodes or elsewhere. Each type of invasion produces its characteristic antibody. Some antibodies have a chemical action, neutralising the poison produced by bacteria; others attack the bacteria directly or make them stick together so as to provide a better target for the roving granulocytes waiting to digest them.

The immune reaction is important for without it we should suffer much more severely from infections. Antibodies continue to circulate even after the attack has passed, providing effective defence which often repels future attacks before they reach unpleasant proportions.

Artificial immunity can be produced by vaccination. Sometimes it is enough to inject a 'killed' vaccine consisting of dead bacteria or viruses; the body's defences recognise the strangers and react by producing antibodies which last for a few years and deal with any further attacks. Live organisms would, of course, lead to more effective production of antibodies, but would also be much more dangerous, especially to a person not previously vaccinated against the disease. A satisfactory response can however be achieved by using a 'live' vaccine made from attenuated bacteria or viruses, that is, organisms which have passed through a succession of laboratory cultures or experimental animals. In the fight against polio during the 1950s, a killed vaccine was developed by Dr Jonas Salk but was superseded by the live vaccine of Dr A. B. Sabin; this vaccine has the advantage that it can be taken by mouth, usually on a lump of sugar.

So far we have looked at two major constituents of the blood, the red cells which convey oxygen to every cell in the body and the white cells which are responsible for defence against infection. The third important component carried by the blood is the platelets. These are formed, like the red and white cells, in the bone marrow. They develop out of large cells but, on reaching maturity, are reduced to small fragments without nuclei. Each cubic millimetre of blood normally contains about 250,000 platelets. They have an important role in the clotting process, which is the main defence against excessive loss in accidental bleeding from a cut or other wound.

Escape from the sea

Having listed the major visible components of blood, we should next look at its chemical composition and function. Blood is a slightly salty liquid and the mineral content is an interesting reflection of our biological history. As we have seen, the earliest and smallest living creatures relied on the sea to bring them oxygen and food and to carry away waste products. As larger and more complicated animals developed, the problem of nutrition became more serious. As long as every cell is in contact with the sea, the supply of oxygen and food (and the removal of waste products) is assured — but the price paid for this assurance is a severe limitation in biological development, since it is necessary for every cell to be within a millimetre or so of the watery environment.

The liquid delivery and disposal system is very suitable for animals; a plant lives in a leisurely way, absorbing its food slowly from air, water, sunlight and soil without any exertion, but an animal does a lot of work in moving about and needs a more concentrated source of energy.

The old-fashioned system, depending on a watery environment, was adapted to changing needs by the development of an internal sea, bathing the tissues of the body. The blood of the early fishes was, in some of its chemical constituents, similar to sea water — but (especially after the development of a pump to drive it round the body) was able to reach cells and tissues remote from the external environment. Eventually the internal circulation, controlled by the heart, became so effective that exchange of materials with the surrounding sea was confined to the passage of oxygen across the gills, the intake of food by mouth and the excretion of waste products from specialised organs such as the kidneys.

The first dry land creatures did not entirely escape from the salty environment, but contrived to carry the sea around with them — internally rather than externally.

Blood is a near relation of sea water and, in one way or another, performs all of the functions of the original medium where life began. The total proportion of mineral salts (mainly sodium chloride) in blood is about 0.9%. Sea water today contains rather more than 1% of salts and the mixture is, in detail, not the same as in blood. It is however, quite plausible to suggest that the mineral content of human blood today is close to that of sea water in the Cambrian period, some 400 million years ago, when life first became established on land.

Our early ocean-dwelling ancestors relied on the oxygen dissolved in sea water, as fish still do. But a fish is a placid creature, untroubled by gravity or by the problem of keeping its body at a higher temperature than the environment. Consequently its energy demands are modest. A man needs a lot of energy and the oxygen dissolved in the water which makes up 80% of the blood would not keep him going for more than five seconds. But anyone can hold his breath for a lot longer than five seconds without harm; where is the rest of the oxygen? It is stored by the haemoglobin of the red cells. Haemoglobin has the useful property of combining freely with oxygen to make a new compound, oxyhaemoglobin, which gives spilt blood its bright red colour. Because of the presence of haemoglobin, the amount of oxygen that can be carried in the blood is more than fifty times greater than the amount that can be dissolved.

The circulation

Having established that it is theoretically possible for the blood to distribute enough oxygen to meet the body's needs, we should now consider the remarkable machinery which drives the circulation.

The specification for this machinery is quite simple. Blood has to be delivered first to the lungs, where it can be recharged with oxygen and relieved of carbon dioxide, then to the rest of the body for the delivery of oxygen, the picking up of carbon dioxide, the redistribution of food and various other rearrangements. Finally the blood must be returned to the heart for its next journey. It might be thought that a single circuit (heart − lungs − rest of body − heart) would be the simplest solution − but in engineering terms this approach presents serious difficulties. For efficient exchange of oxygen and carbon dioxide in the lungs, the blood must be close to the surfaces over which the inspired air is passing. A large blood vessel is of no use in this connection. As we have already seen, diffusion takes place at a reasonable rate only over very short distances; it is therefore necessary for the blood to be flowing through the lungs in very small vessels, usually called capillaries, so that all of it can be close to the inspired air for exchange purposes.

The passage of blood from a large vessel (leaving the heart) into a multitude of small vessels produces some technical problems. Friction is greater in a narrow tube than in a wide one, simply because a greater proportion of the flowing liquid is near the walls, where the drag is greatest. Since the blood flows more slowly through the capillaries than through the artery which supplies them, the total cross-sectional area of the capillaries must be greater than that of the artery, for otherwise there would be a pile-up of blood. In fact the problem is even more serious in the circulation of blood than it would be for a network of water pipes. The capillaries are so narrow that only one red cell can pass at a time. In this situation the cells rub against the walls and add to the frictional drag.

The capillaries are however designed to cope with the engineering requirements, and their total cross-sectional area is many times greater than that of the artery which feeds them. Another complication now becomes apparent, for the pressure in the capillaries is, because of the increased area, lower than in the original artery. This is a simple mechanical effect; a river which thunders through a narrow gorge will flow quite calmly again when it reaches a wider channel.

If the blood, after passing through the lung capillaries at low pressure, had to be collected up again into wide arteries for efficient transport to further parts of the body, a lot of additional pumping would be necessary, representing a considerable load on the heart.

The solution adopted in the human body is simple and effective, with the heart driving blood through two separate circuits. In the first circuit, blood is pumped from the right side of the heart through the pulmonary arteries, which subdivide into smaller vessels (the arterioles) and then into a network of fine capillaries from which the exchange of oxygen and carbon dioxide can be effected.

When we breathe in, air passes through the trachea

(or windpipe) into the lungs, where it is distributed in a multitude of branching pathways which lead to smaller and smaller passages eventually terminating in millions of tiny air sacs or alveoli. The multiplicity of airways gives the lungs a spongy structure with an enormous surface area for gas exchange; the total area of all the alveoli, if they were stretched out flat, would be about a thousand square feet.

The blood which comes from the heart into the capillaries of the lung has lost most of its oxygen (in previous passage round the tissues of the body) and has picked up a good deal of carbon dioxide. Consequently the fresh air on one side of the alveoli is considerably richer in oxygen than the blood flowing through the capillaries on the other side. In these circumstances oxygen diffuses through the walls of the alveoli, through the walls of the capillaries and into the blood where it is promptly captured by haemoglobin molecules. In the same way the blood flowing through the capillaries is richer in carbon dioxide than the fresh air on the other side of the alveolar wall. Consequently carbon dioxide passes out of the blood, across the alveolar wall and into the lungs. As a result of these exchanges, the composition of the air and the lungs is changed appreciably. We breathe in fresh air containing about 21% oxygen and 0.03% carbon dioxide but the air which we breathe out contains about 16 or 17% oxygen and 3.5% carbon dioxide; the remaining 79 or 80% in each instance is made up of nitrogen.

Purged of carbon dioxide and replenished with oxygen, the blood is gathered up in the pulmonary veins and returned to the left side of the heart. It arrives first in the left auricle (or atrium), a thin-walled cavity in the heart muscle. The auricle contracts and sends blood through a flap valve into the left ventricle, another cavity in the heart muscle with quite thick walls. The ventricle is not very large and the distention of its walls by the incoming blood generates a considerable pressure. When the ventricle contracts the blood does not flow into the auricle (because the flap

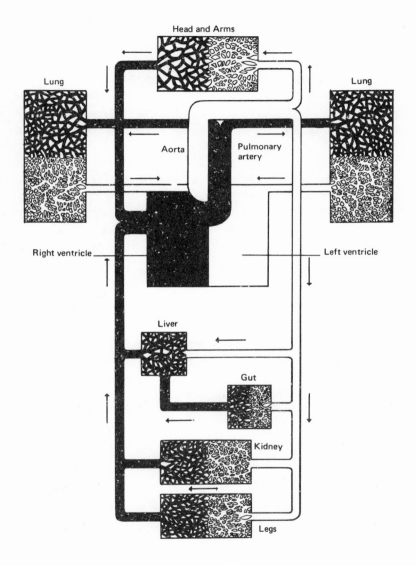

Figure 6 diagrammatic sketch of the circulation

valve opens only in one direction) but is pushed out through the aorta, a large artery which subdivides to supply the rest of the body.

The intermittent action of the heart, with one powerful beat every second or so, is not at first sight well adapted to the needs of the other organs and tissues for a continuous supply of oxygen and other materials carried by the blood. In engineering terms, an intermittent pumping action is very inefficient, since the pump, pipes and valves have to be built to withstand the full pressures and flow rates, though these conditions exist for only a small part of the pumping cycle.

The stop-go action of the heart is however converted into something more like a continuous process by an ingenious mechanism, illustrating sound technical principles. The engineer faced with the problem of smoothing out an intermittent supply would probably think of an expansion chamber to collect surplus liquid at the start of the pumping cycle and release it during the interval between successive strokes. He might even think of a regulator, consisting of a section of pipe with elastic walls, which would stretch during the pumping stroke. When the force of the pump reduced, the wall of the regulator section would return to its normal position, releasing liquid into the circuit.

The body has a regulator rather like this, because the large arteries have a considerable amount of elastic tissue in their walls and are able to stretch when put under pressure. The blood rushing out of the left ventricle into the aorta does not at first go very far because a lot of it is accommodated by the stretching of the aorta. When the contraction of the ventricle has finished, the aorta returns to its normal state and releases blood into the smaller arteries to maintain a steady supply.

The pulse beat that can be felt in the wrist and at other places in the body (wherever an artery lies just under the skin) is a remnant of the shock wave produced by blood spurting out of the heart. Though the force of the disturbance is, as we have just seen, smoothed out by the elasticity of the

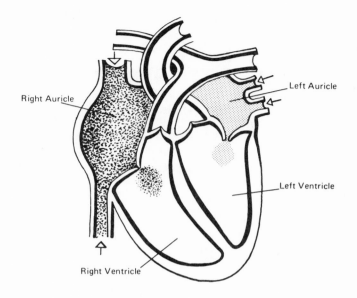

Figure 7 diagram of the heart

arteries, it can be demonstrated quite easily by sitting with the legs crossed; as blood flows into the upper leg at each heartbeat the foot reproduces the pulse wave in jerky movements of a millimetre or two.

Once free from the heart, the blood is distributed first into arteries, next into smaller vessels known as arterioles and then into narrow capillaries, where it exchanges oxygen and carbon dioxide with the tissues of the body.

On leaving the capillaries the blood flows into the venules and finally into the veins which lead it back to the right auricle, from which it passes to the right ventricle and, as we have already seen, out into the lungs. Blood which has been supplied to the digestive tract is collected in a separate system of veins leading to the liver, where food molecules are removed for chemical processing before the blood is returned to the heart. Another loop of the circulation takes arterial blood to the kidneys, where toxic substances and other waste products are removed and the concentration of various mineral salts is adjusted.

Pumping not enough

It is often supposed that the heart pumps the blood round the body – but this is a misleading simplification. The heart is a well-engineered machine which discharges blood into the arterial system with enough pressure to take it to its destination, but not much more. The pulmonary circulation, supplying the lungs, operates at a rather low pressure just enough to take the blood, against the force of gravity, to the apex of the lungs, a vertical distance of about fifteen centimetres. In the aorta, near the heart, the maximum arterial pressure is, for a normal person, 120-150 mm or about 2½ lbs per sq in. Nearly all of this pressure is lost by the time the blood arrives at the extremities; how then does it return to the heart? Gravity takes it back from the head and other parts of the body above the level of the heart, but the return from the feet is a more difficult problem.

Blood makes its way to the feet by the combined effects of the heart pump and gravity. Having reached its lowest point, the blood has very little pressure left to take it on the long uphill journey back to the heart. The veins are however supplied at fairly frequent intervals with valves which allow blood to flow towards the heart but not in the reverse direction. Consequently any pressure exerted by the surrounding muscles will cause the blood to move upwards.

In the normal way, the movements of the muscles in walking about, or even when sitting, are enough to squeeze blood through the veins. A person standing still for a long time may not give the blood enough help. If the return of blood to the heart is impeded, the supply of blood to the brain also suffers. The faint is nature's way of dealing with this situation by bringing the victim to a horizontal position where gravity is no longer a nuisance.

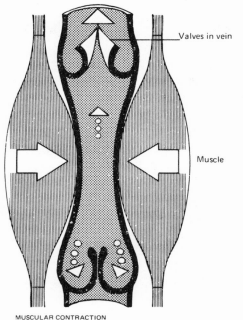

Valves in vein

Muscle

MUSCULAR CONTRACTION
helps the upward flow of blood from the feet and legs

Figure 8 muscular contraction

The return of blood through the veins to the heart is also assisted in a natural way every time we breathe. Air, like blood or any other fluid, only moves in response to a difference of pressure. When we breathe in, the pressure in the lungs must be reduced below that of the outside air and when we breathe out the pressure in the lungs must be higher than atmosphere — otherwise the air would not move.

Air is drawn into the lungs by the lowering of the diaphragm, the sheet of tissue which separates the chest cavity from the abdomen. When the diaphragm is lowered (by signals from the brain to the appropriate muscles), the rib cage swings outwards and upwards, also under automatic control; all of these actions cause the volume of the chest cavity to increase.

When the volume of a quantity of gas is reduced (for example in pressing on a tyre pump) its pressure is increased — that is how we force air into the tyre. Conversely, if the volume of a quantity of gas is increased, its pressure is reduced. This is what happens in the chest cavity when the diaphragm falls. Since the pressure in the chest (including the lungs) is less than the atmospheric pressure outside, the lungs are able to suck in fresh air; the fall in pressure also applies to the heart and the veins inside the chest cavity, which are accordingly better able to suck in blood from the veins below the diaphragm. The effect is further improved because the movement of the diaphragm which enlarges the chest cavity compresses the abdominal cavity and consequently increases the pressure there. The veins are therefore compressed (just as they would be by voluntary muscular action) and the ascent of blood to the heart is further aided.

After this brief summary of structure and function, we may attempt an engineering appraisal of the heart, considering it first as a pump. In this regard, its most important design feature is that it does not deliver liquid continually. Many circulating pumps have a rotary action. The water pump of a car, for example, has a miniature paddle-wheel, driven by the engine. Cooling water delivered near the centre of the wheel is trapped by the vanes and spun to a high speed before reaching the delivery pipe from which it flows to the cylinder block.

The heart has no rotating parts but its design is rather like that of the fuel pump in a car. This is usually a diaphragm pump, in which a reservoir is connected to the petrol tank and closed at its other end by a springy metal disc which moves backwards and forwards. The back stroke enlarges the reservoir, which then fills with petrol; the succeeding forward stroke pushes the petrol through a pipe to the carburettor. The two pipes, from the tank and the carburettor, are fitted with valves to prevent flow of fuel in the wrong direction.

The heart really consists of two pumps, one supplying the

lungs and one supplying the rest of the body. The duty cycle of the heart muscle is arranged so that the two pumps operate in the proper time sequence and the mechanical design allows for the fact that the left side of the heart, supplying the greater part of the body, has to do six or eight times as much work as the right side which drives the low pressure circulation of the lungs. The left ventricle contains about twice as much muscle as the right and therefore has to work three or four times as hard.

Efficiency of the heart

When an engineer has designed a machine he is interested in knowing its efficiency — that is, the amount of useful work that it does in relation to the energy that it uses in the form of fuel. The work done by the heart can be estimated from a knowledge of the amount of blood pumped out at each stroke and the pressure and speed at which it leaves. The energy supplied to the heart can be found by measuring its oxygen consumption, usually by analysis of blood samples obtained in cardiac catheterisation, a technique in which a thin tube is passed into the heart. Tests of this kind indicate an efficiency of about 40%. For comparison, the efficiency of a steam engine seldom exceeds 20%. Bearing in mind the small size of the heart (it weighs less than twelve ounces) and the ease with which it adapts to widely varying demand, the efficiency of 40% is an excellent achievement.

As a machine, the heart has to be quickly adaptable to varying loads. When sleeping, the energy expenditure of the body is about 0.025 hp. In gentle walking, the energy expenditure corresponds to about 0.1 hp. A sprinter can work at 1 hp or more for a short time and a champion long-distance runner can achieve an output of 0.4 to 0.5 hp for several hours.

In adaptation to these widely varying demands, two processes are important. Firstly the heart rate is greatly increased, to as much as 180 or 200 per minute, about three times the normal value. The breathing rate, normally about fifteen per minute, is increased in about the same proportion and the depth of breathing — that is, the amount of fresh air taken into the lungs at each breath — also rises. The amount of blood pumped out by each heartbeat increases only slightly and, in severe exercise with a greatly increased heart rate, may actually diminish because there is not enough time for the heart to fill with blood between successive beats. The blood pressure, however, rises so that the blood supply to the muscles is improved.

Temperature regulation

The animal kingdom is divided into cold-blooded creatures (including birds, fish and insects) and warm-blooded creatures, including man and other mammals. It would be more correct to say that warm-blooded animals keep a constant internal temperature — about 37°C in man, 39°C in other mammals and 42°C in birds — while the others allow their internal temperatures to fluctuate with the environment. A man (or a bird) can travel from the Arctic to the tropics with very little change in body temperature. A frog will not survive severe sub-zero temperatures but otherwise will (to within 2-3 degrees) take up the temperature of the water in which it is put. Some animals, such as the dormouse and the hedgehog, are warm-blooded for most of the year but become cold-blooded for their winter hibernation. Hibernation begins when the ambient temperature drops below about 12°C; in technical terms, the body's thermostat is then switched off.

In man and most other mammals the regulation of temperature is very efficient; the blood is important in this process for three reasons — because it circulates, because of its high specific heat and because of its high latent heat.

In some ways the circulating blood acts like the cooling water of a car which takes heat from the engine, delivers some of it to warm the passengers and loses the rest by radiation or convection into the air. The blood, to mention one example, picks up heat from the liver (which, because it is the scene of many vigorous chemical actions, stays at least 1°C warmer than the rest of the body) and gives it up to the skin. On a hot day, or during vigorous exercise, the skin acts like a car radiator and loses heat to the air. In these circumstances, the capillaries near the skin open wider, allowing more blood to flow and more heat to be brought to the surface. On a very cold day the capillaries contract in parts of the body covered by clothing, but expand in the exposed parts (such as face and hands), so delivering heat to reduce the risk of frostbite.

Blood is 80% water and water is a good medium for warming or cooling because of its high specific heat, the highest of any common and harmless liquid. This means that, for a given change in temperature, it will pick up (or give away) more heat than any alternative liquid.

Another useful property of water in this connection is its latent heat which is also very high. Water exposed to the air will evaporate, slowly on a cool day and more quickly on a warm day. Quite a lot of energy is needed to change from the liquid to the vapour; this energy is called the latent heat. When we boil an electric kettle, the energy is supplied from the mains, but in other circumstances the energy is taken willy-nilly from the immediate surroundings.

A damp cloth applied to a feverish forehead brings relief, for the water removes quite a lot of heat from the body as it evaporates; an icebag has much the same effect, since heat is also needed to turn ice into water. In normal circumstances more than a pound of water evaporates from the skin every day in non-obvious perspiration. The latent heat accompanying this loss is more than 300 kcal or roughly 10% of the daily energy intake provided by the diet. The heat loss from a body wrapped in wet clothing can be much more serious and is sometimes the cause of death after shipwreck.

In warm weather the ability to augment the body's heat loss is very useful. On a hot day the circulating blood gives up quite a lot of water which appears on the skin as sweat and, in evaporating, takes away heat. Perspiration is not always apparent; the skin only becomes moist when the rate of production of sweat exceeds the rate of evaporation.

There are two kinds of sweat, both produced by glands in the skin. The major source is represented by the eccrine glands, of which there are two million in man; they occur all over the body but are most numerous on the forehead, cheeks, hands and feet. The sweat that they produce is almost pure water, with a little salt; in extreme conditions of heat or exertion the salt loss can be troublesome, leading to muscle cramp if not made good. During hard exertion the production of sweat can be a litre per hour, corresponding to a salt loss of three grams, or a third of a normal day's intake. Eccrine sweat has no smell, but the body also possesses apocrine sweat glands, concentrated in the armpits and in the genital regions. These glands produce sweat which is of little value in cooling the body (since the regions concerned are not in contact with the air) but contains organic substances which, when decomposed by the bacteria always present on the skin, produce unpleasant odours.

Fat people perspire more than thin ones, for several reasons. Having a larger surface area, they have more sweat glands. Because of their greater weight, they produce more heat in the course of exercise or other physical activity. They are also more conscious of perspiring because some of the sweat is trapped by folds of skin and cannot easily escape. Since obesity is, for most people, a voluntary affliction, the fault lies with the user and not with the design of the cooling system.

Notwithstanding the ingenious arrangements which have been mentioned for conserving heat, the temperature of the circulating blood can sometimes drop quite substantially. For a long time it was thought that the temperature of arterial blood differed little from that of the central core of the body – but recent experiments have shown that blood in the radial artery in the arm can fall to 21.5°C while the central body temperature remains at the normal 37°C.

This state of affairs was explained by the discovery that the body uses the counter-current exchange system already familiar in the chemical industry. Briefly, this technique achieves more effective transfer of heat by having the hot and cold streams flowing in opposite directions and in close contact. The main arteries in the arm are sandwiched between veins, allowing very favourable conditions for heat exchange. On a cold day, arterial blood reaches the forearm, near the elbow, at a temperature which may not be much below 37°C. The veins on either side of the typical artery are, at that point, carrying blood returning from the hand and not yet restored to its normal temperature. This blood is therefore warmed by the arterial flow. The arterial blood moves on, somewhat cooled, but the adjacent veins contain blood which is even cooler and therefore still capable of absorbing heat.

The same situation prevails down the whole length of the artery and its closely adjacent veins. Consequently heat is effectively transferred and, more important, is conserved in the circulating blood. Were the arteries and veins not so conveniently close together, the arterial blood would lose appreciable amounts of heat to the surrounding tissues in its passage down the arm and the returning venous blood, at a relatively low temperature, would absorb a further amount of heat on reaching the warm core of the body.

On a really warm day, the heat-conserving mechanism just described is not so useful, since it prevents loss of heat to the outside air. However, the arm has an alternative network of veins just under the skin; in hot conditions, the venous blood is diverted to this system.

Repairing the puncture

The important properties of blood all depend on the fact that it is continually on the move. Since the blood is everywhere under pressure (otherwise it would not move) it is bound to spill, possibly quite vigorously, wherever a leak occurs. The body contains no more than six litres of blood; half a litre can be given to the blood bank without discomfort but loss of a litre would produce noticeable effects and loss of two litres can be quite dangerous. This represents more than a third of the oxygen-carrying capacity; without it the brain and other vital organs are in jeopardy.

Fortunately the circulatory system is the original self-sealing tank, with a complicated but successful mechanism to deal with all but the most severe accidents involving blood loss.

The basic principle of the self-sealing technique is that the blood, when it comes in contact with the outside air, quickly solidifies in a clot, which holds back further bleeding and acts as a temporary patch until the underlying tissues have been repaired. The mechanism is complicated because, although the sealing action is required quickly in an emergency, the formation of a clot in the circulating blood can be disastrous, as for example in coronary thrombosis.

The blood clot which forms over a cut is a tangled mass of fibres with blood cells trapped between them. The fibrous material is a protein known as fibrin. Fibrin however cannot be present in the circulating blood, since otherwise it would clot throughout the body. The presence of a cut or the exposure of fresh blood to the outside air causes the production of fibrin from another protein called fibrinogen, which circulates freely in the blood. The substance which converts fibrinogen to fibrin is thrombin.

At this point the argument begins to appear endless because, clearly, if there is any thrombin in the circulating blood it will convert fibrinogen to fibrin and cause clotting everywhere. The full story of the clotting mechanism is complicated and probably not yet fully explored. It is, however, known that the key to the mechanism resides in materials released by the breakup of blood platelets. After that comes an elaborate sequence of checks and counter-checks with no less than thirteen clotting factors including calcium (always present in the blood) and fibrinogen. Absence of clotting factor VIII causes haemophilia A, the common form of the disease, and people without clotting factor IX suffer from haemophilia B — sometimes known as Christmas disease, after the surname of the first victim described in the medical literature.

The ability of blood to form a protective clot when exposed to the air is an important first aid measure, but is a nuisance when a sample of blood has to be taken for examination since, if something is not done to modify the process, it will be firmly clotted before the pathologist or the biochemist has a chance to look at it. A simple precaution is to collect the blood in a bottle containing heparin — a substance which occurs in the body, particularly in the liver and lungs, and is one of the defences against internal clotting. An even simpler method is to put a little citrate or oxalate solution into the bottle when collecting blood. Calcium ions (clotting factor IV) are swiftly removed by citrate or oxalate ions and the clotting mechanism is therefore put out of action.

The clotting mechanism is complicated but it offers an effective solution to a difficult problem. The engineer recognising the possibility of leakage in a pipework system would think first of replacing or patching the appropriate section. Solutions of this kind are of little use in repairing the human machine because the natural materials are unique and are seldom compatable with non-biological substitutes. An engineer would certainly be delighted to work with a liquid which, flowing in an elaborate network, would not only detect a leak but would, without any additional materials, effect an immediate repair, tidy up the debris and leave the damaged part as good as new.

The self-sealing power is only one of the remarkable properties of blood. As we have seen, this familiar liquid provides the cells and tissues of the body with self-service facilities for their varied needs in food and waste disposal. It also provides an excellent control mechanism to keep the body temperature constant. Furthermore, it never wears out but continually renews itself. It is not surprising that blood has a significant role in most religions, for it has rightly been named the river of life.

9: the inner ocean

Though the body is to outward appearances a solid structure, its inward arrangement still reflects our marine ancestry. For a start, the body tissues contain a great amount of water — 50 to 70% of the total mass. The proportion cannot be given more precisely, because the fat which we all possess to a greater or lesser extent contains practically no water. Consequently a very thin person may be almost 70% water while in an average man the proportion is about 58%.

In man and many other animals the adaptation to life on land is evident in the complicated and successful mechanisms which have evolved for the conservation of water. These mechanisms are not of much importance for simple creatures living in the sea but in man they are of impressive efficiency.

In power stations and other factories the great amounts of water used for cooling or chemical processing are bought at little or no cost and, after use, discarded into the sea, the river or the sewers. The works manager or chemical engineer ordered to operate the plant so as to use every drop of water a hundred times would hardly know where to start — but this is the extent to which body water is conserved.

In considering the water balance of the body, two important requirements have to be met. Firstly it is important to keep the fluid volume of the body constant within very narrow limits. Water loss amounting to 1% of the body weight produces distinct thirst. Loss of fluid equivalent to 5% of the body weight results in collapse and a 10% loss is fatal. A man can last for weeks without food but only for a very few days without water.

Secondly it is important that the chemical composition of the blood must be scrupulously maintained; for some important constituents, such as sodium and calcium, a departure of 10% from the normal level can have alarming consequences. It is necessary also to have an efficient system for the removal of waste products which accumulate during the digestive processes, since some of these materials are quite toxic even in small amounts.

The blood is really a liquid assembly line. Hardly any of its products are exported, because virtually the whole of its productive capacity is needed for the maintenance, repair and extension of the factory, that is, the body tissues. The system presents difficult design problems in chemical engineering.

Consider for example the disposal of waste products. The body, like any other engine, burns its fuel. True, the burning is not accompanied by flame or explosion — but the foods that we take in to fuel the body are oxidised in equally effective if less spectacular processes. Fats and carbohydrates are not too difficult in this regard. Since they are composed entirely of carbon, hydrogen and oxygen, the oxidation products are carbon dioxide and water, which can be dealt with easily by adding the water to the blood and by adding the carbon dioxide to the expired air from the lungs. Unfortunately we cannot live on fat and carbohydrate, though many people enjoy a diet which consists largely of these two components.

The trouble is that every piece of living matter is made of protein. Since the cells of the body are in a continual state of breakdown and removal, an appreciable amount of fresh protein must be supplied to make up the loss. Some of this loss, in skin, hair and nail, is fairly obvious, but blood cells and a number of other tissues are also continually renewed. All proteins are not equal; the sheep or pig protein that we eat is not a direct replacement for human protein. Indeed, if a scrap of any of these foreign proteins found its way inside the body, it would be immediately rejected by the process that has already been described (page 117).

For these reasons it is necessary for meat and other foreign protein to be broken down chemically before it enters the body. In this regard, the stomach and other parts of the digestive tract are, of course, outside the body since they represent a tunnel continuous with the external skin and are not, strictly speaking, part of the interior of the body at all.

The proteins taken in food are broken down into their constituent amino acids which are then reassembled inside the body to make human protein of various kinds.

There is naturally a certain amount of waste in this process, and an important part of the waste from protein contains nitrogen. The disposal of nitrogen from the body is a difficult chemical problem. It might be thought that the problem could be dealt with by merely exhaling nitrogen gas. Unfortunately a good deal of chemical energy would be needed to turn the nitrogenous protein waste into nitrogen gas. Quite apart from this difficulty, nitrogen is not very soluble in water or in blood.

The bends, a painful and sometimes fatal condition produced in divers who return to the surface too quickly, is brought about because, when the external pressure is reduced (on coming to the surface) nitrogen gas is released from the blood or from fatty tissue in the form of bubbles. As the external pressure drops on nearing the surface, these bubbles expand, causing damage to the tissues or blocking blood vessels. (Oxygen is released from the blood at the same time but is quickly used up in the normal processes of the body and does not constitute such a serious hazard.)

Another possibility would be to turn the nitrogenous protein waste into ammonia; this is a very soluble gas and its production in the body needs little or no energy. Unfortunately ammonia is poisonous to most forms of life. Tadpoles and some fishes dispose of their nitrogenous waste in this way but are able to release it immediately to the watery external environment, where it is diluted to a harmless level.

In man a better solution has evolved; the ammonia, which

is the first breakdown product of amino acids, is swiftly combined with carbon dioxide to form urea, a soluble compound. A diet with an adequate protein content would produce, every day, twenty or thirty grams of urea, which cannot be allowed to accumulate in the blood but must be discarded from the body. The vehicle for this removal process is the urine, a fluid produced by the kidneys.

The urine is a pale shadow of the blood. The kidneys are responsible for the chemical transformations involved in this change; an idea of the specification to which a kidney works can be gathered by looking at the differences between raw material and final product. Some substances are concentrated very effectively in the urine. There is for example a small amount of ammonia in the circulating blood, but the concentration in the urine is five hundred times greater. Other materials, such as proteins, amino acids, vitamins and glucose, are quite easily detected in the blood but are almost completely absent from the urine. For yet another group of substances, such as sodium and calcium, the concentrations in blood and urine are much the same.

The kidney is therefore not merely a simple filter or concentrating device. In fact, it monitors the concentrations of at least thirty chemicals in the blood, removing some, conserving some and adjusting the levels of others so that the blood can carry out its various tasks efficiently. The kidney also helps to maintain the acidity of the blood at a correct level and to stabilise the blood pressure and the blood volume.

A chemical engineer, faced by this formidable specification, would realise immediately that the task could not be performed as a single operation; he would make a flow chart, involving a succession of specialised pieces of equipment for the different stages in the processing. The kidney is not very large, but the available space is ingeniously used. For a start, the blood flows through it in more than a million narrow tubes, rather than in one large vessel. This design provides a much bigger surface area for a transfer of chemicals from vessels carrying blood to those collecting urine. The chemical transfer is a slow process, since it has to be effected at normal body temperature without using any materials apart from those present in the blood and other tissues and without any mechanical aids such as the pumps and centrifuges that the chemist uses in the laboratory. For these reasons it is necessary to have a great length of tubing, actually about thirty miles, in each kidney.

The design also has to allow for the fact that the blood is not a suitable material for passing through very fine tubes, where the red cells would cause blockage.

All of these difficulties are dealt with quite successfully in the kidney. The chemical processing takes place in two main stages. Firstly the blood is filtered, leaving the cells and the plasma proteins (p.116) behind but removing most of the water and dissolved solids. In the second stage the watery filtrate is processed further, so that the useful materials (such as glucose and vitamins) which it contains are put back into the blood, while the waste products and other surplus materials are carried away.

The processing starts when blood arrives at the kidney (through the renal artery) and is immediately distributed to about a million tiny filter units, where the watery component is removed. The filtration is greatly helped by the fact that the renal artery is quite a short tube, connected to the aorta, where blood from the heart is still at a high pressure; much of this pressure is available for forcing the water and dissolved solids of the blood through the tiny spaces in the filter units.

The filtered fluid now passes along the narrow tubules, which are enmeshed in blood capillaries connected eventually to the renal veins. Most of the fluid is in fact re-absorbed into the blood through these connections; the combined output from the filter units in the two kidneys is about 130 ml per minute but the rate at which urine accumulates is only about one ml per minute. In other words, more than 99% of the liquid filtered out of the blood is re-absorbed into the circulation.

Chemical pumps

How are the water and dissolved chemicals put back into the blood? There is no simple answer to this question. Since, for example, the concentration of sodium is much the same in the filtrate as in the blood, there is no obvious reason why any of it should make the return journey. The transfer is effected by chemical pumps. Specialised cells in the tubules use energy (derived, in the last resort, from food that we eat) to push sodium ions back into the blood capillaries which are in close contact with the tubules at a great many places. The sodium ions are positively charged and, when a good many of them have made the crossing, their accumulated positive charge attracts negatively charged ions (such as chlorine) from the filtrate into the blood. The transfer of sodium chloride in this way makes the passing blood more concentrated and thereby promotes the movement of water through

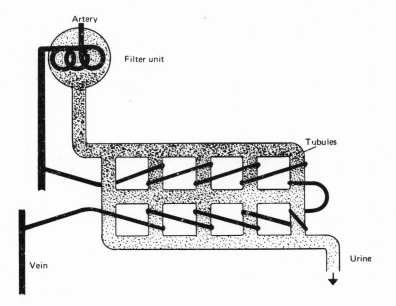

Figure 9 diagram showing action of the kidney

the wall of the tubule into the blood capillary.

A different kind of pump is responsible for the recovery of glucose, which is a valuable food. This pump can be overloaded, for example, by eating a lot of sugary food, and in that event some of the glucose escapes into the urine, because the pump does not have enough capacity to push it back into the blood. In normal health, surplus glucose circulating in the blood is converted to glycogen, which can be stored in the liver. This conversion requires insulin, a hormone usually present in the blood. In a person without an adequate supply of insulin, the pump in the kidney is greatly overloaded and much of the glucose is lost in the urine. Its presence there is easily detected and forms the basis of a simple test for diabetes, a disease which is dealt with by administration of insulin to restore the normal mechanism for dealing with glucose in the blood.

In one way and another, the materials required in the blood are restored in the process of tubular re-absorption, leaving the unwanted chemicals and a relatively small amount of water to flow down the ureters into the bladder.

The bladder can hold about a pint of urine, but discomfort usually begins when it is half full. At this stage, if not before, nerve endings in the bladder wall generate a signal corresponding to the increased pressure. The brain responds by commanding a contraction of the bladder to expel the urine. In an infant, this process occurs without hindrance, but a child soon learns to control the message from the brain, so that the release of urine can be carried out in a socially acceptable manner.

In engineering terms, the bladder is a self-emptying cistern, controlled by a strain gauge which triggers the release mechanism when the internal pressure rises to a preset level. Normally the system works very well, but it can go wrong for a variety of reasons. In old age, for example, the muscles which normally guard the exit of the bladder may go out of control, producing urinary incontinence, a distressing affliction which is more common than might be supposed. In younger people, the strain gauge in the bladder wall may be disconnected from the brain by an injury to the spinal cord. Modern electronic technology offers a way of dealing with these difficulties, which will be described in a later chapter.

Urine, providing an easily available sample of the internal environment, has always been treated with respect by physicians. In earlier times, whole systems of medical diagnosis and treatment were based on its appearance and chemical analysis.

Our ancestors, living in frugal times, found interesting uses for material which might otherwise have been wasted. Stale urine produces considerable amounts of ammonia, which served as a cleaning material in the days before soap, and is still used in this way by the Eskimos. The Emperor Vespasian (who reigned from AD 70–79) made great improvements in the Roman health service and was one of the first to pay doctors' salaries out of public funds. The cost of these innovations did not worry him much for he displayed a bizarre fiscal ingenuity. His most notable achievement was the lavish provision of outdoor toilet facilities with storage tanks for the urine, which he sold to the launderers, thereby advancing the public sanitation while enriching the imperial coffers. This notable financial enterprise was long commemorated by the vespasiennes which, until quite recently, stood in the streets of Paris for the public convenience. Vespasian's son, afterwards the emperor Titus, chided him for raising money by such unseemly methods; in reply he held up a bag of gold and uttered the celebrated Latin tag: 'non redolet' (meaning that the money did not smell).

Water conservation

Automatic adjustment of the rate of urine formation provides an ingenious and sensitive method of controlling both the volume and the concentration (that is, the content of salt and other minerals) of the blood. The control system is a negative feedback loop, a device which uses the same basic principle as the thermostat found in refrigerators, central heating systems and many other domestic situations. In an immersion heater used for hot water supplies, a temperature sensing element switches off the current when the temperature rises above a certain value and switches it on again when the temperature falls one or two degrees below the required level.

There are no mechanical switches or large electric currents in the body, but many negative feedback systems use chemical messengers called hormones. For controlling the concentration and volume of the blood, the messenger is anti-diuretic hormone or ADH. The concentration of this substance in the blood is continually monitored by specialised cells in the brain. If the blood becomes more concentrated, for example after loss of water by perspiration in vigorous exercise, the monitoring cells generate an electrical signal which is conveyed to the pituitary, a small gland in the brain, where ADH is released into the bloodstream. The hormone travels to the kidneys and, in a way which is not fully understood, reduces the formation of urine, thereby conserving the body's water supply and allowing future fluid intake to be used for diluting the blood back to a normal concentration. If the appropriate part of the pituitary gland is put out of action by disease or (in experimental animals) by cutting the appropriate nerve, the output of urine can increase by as much as fifteen times above normal; automatic adjustment of the ADH output from the pituitary gland is therefore a very sensitive method of stabilising the concentration of the blood.

There is no obvious way in which the body can measure its own blood volume, but there are pressure-sensitive cells in the large arteries and cells sensitive to stretching in the major veins as well as in the auricles of the heart. When the blood volume increases, the veins and auricles are stretched and the monitoring cells generate electrical signals which pass through the nervous connections to the brain, leading to a reduction in the flow of ADH and consequently an increase in the production of urine, thereby helping to reduce the blood volume. The pressure monitoring cells in the large arteries act in a similar way.

Alcohol, the most socially acceptable poison, interferes with many of the chemical systems of the body. In particular it slows down the production of ADH and, for this reason, increases the output of urine. For every ounce of alcohol drunk, ten ounces of extra urine are produced. The diuretic effect of beer is well known, but is a consequence of the water content rather than the alcohol, which makes up only three of four per cent of the drink. (Alcohol, incidently, has a very high calorific value – which is one reason why beer drinkers usually become fat.) Strong drinks, such as gin, brandy and whisky, where the alcohol content may be as much as 45%, do not, if taken neat, do anything to help the thirst; quite the reverse, since the consumption of one ounce of whisky (containing, at the most, half an ounce of water) will deplete the body of a further four and a half ounces of water in the form of urine. Wine, with an alcohol content of 10 to 12%, is the most satisfactory form of indulgence – as judged by the physiologist – since the eventual excretion of urine is just about equal to the total volume drunk.

A somewhat similar problem is faced by the shipwrecked mariner who has nothing but brine to drink. The mineral content (mainly sodium chloride) of sea water is about 3.5%, but the human kidney is not capable of making urine with salt concentration of more than 2%. Consequently there is a net loss of water from the body after drinking sea water – leaving the victim thirstier than ever.

For a fish, the problem of maintaining the correct amount of body water is even more serious than for a man. A fresh water fish has difficulty in keeping the mineral content of its blood at the proper level. One of the commonest biological processes is the movement of water through the boundary separating a dilute solution from a concentrated solution. A fresh water fish is at great hazard because of the natural tendency of the surrounding water to pass through the skin, so as to make the concentrations more nearly equal inside and outside. A fish's skin is of course covered with scales which act as effective waterproofing. On the other hand, part of the fish's exterior must allow the passage of small molecules, since that is the only way in which oxygen can be absorbed for the support of all the vital processes. Consequently some of the water which flows over the gills inevitably passes into circulating blood and has to be removed by the kidneys which are relatively large.

In this situation, the familiar warning about drinking like a fish is not valid — a fresh water fish has to restrict its water intake severely and drinks very little. For a salt water fish, however, the problem is quite different. The proverb explaining that blood is thicker than water is true for man but not for a sea fish; there the water is thicker than blood and the danger is that water will be lost from the circulation in vain attempts to dilute the concentrated sea water flowing over the gills. A salt water fish has to drink heavily to conserve the fluid balance. We have already seen that a man who drinks sea water will be in dire straits because of the subsequent excessive loss of urine. A salt water fish is not able to produce concentrated urine but is able to push the salt out again through the gills.

The human animal enjoys immersion in fresh water and in salt water. Fortunately our skin is adequately waterproof — were this not so we should be in danger of shrinking in the sea and bursting in the bath.

Reviewing the complex activities of the kidney, the biologist sees this organ as a major evolutionary development, linking our internal environment (still wet and salty) with the outside world, now mainly dry. The engineer sees the kidney as a versatile chemical processing plant, dealing successfully with a material which can vary considerably in composition and in quantity. A man can live on little more than a pint of water a day (in addition to the water present in food) but can drink sixty pints in quick succession without overloading the kidney.

The design of the kidney does however have one surprising and potentially dangerous feature. A chemist wishing to purify a liquid would probably find ways of removing the unwanted substances one by one; the kidney discards almost everything and reclaims the essential contents from the waste pipe. The leakage of vital chemicals from the blood could be disastrous, but the reclamation seldom goes wrong; even when the kidneys fail, the consequence is usually an accumulation of poisons in the blood rather than a loss of valuable substances.

Common experience in the domestic and industrial worlds suggest that the kidney, with its miles of fine-bore tubing, should have trouble from furring of the pipes. Most of what passes through the tubules is a simple solution – but the urine does contain materials such as uric acid, calcium phosphate and calcium oxalate, which do not dissolve but remain as tiny crystals.

In normal conditions the microscopic fragments are carried along and discharged from the body without difficulty – but two questions remain: why do the crystals not stick to the walls of the tubules and why do they not stick to one another to form large lumps?

The answer to the second question is that they do sometimes stick together to form stones which, jammed in the kidney or the ureter, are very painful. 'Cutting for the stone' was a familiar surgical achievement in earlier times, when the condition was much more common than today, perhaps because of changes in diet.

Uric acid is always present in the blood and forms compounds which sometimes gather, also as crystals, in the joints of the head or foot, particularly in the big toe, causing gout. This affliction is traditionally associated with over-indulgence in rich food and drink, but its origin is not well understood; the fact that it is hardly ever found in women suggests that some genetic factor may be involved.

As to how the microscopic crystals of phosphate, oxalate or uric acid are prevented from clumping together to form stones in the normal kidney, we do not know; the answer to this problem would be welcomed, not only in medicine but also in industry, where furred pipes cause heavy financial loss.

The kidney displays a mastery of chemical engineering and fluid mechanics which the technologist envies but cannot emulate. Next to the brain, it is the most complicated — and the most reliable — organ in the body.

10: food and fuel

From time to time newspapers carry the story of an inventor who knows how to make a car run on water. Demonstrations are sometimes given and it is not difficult to attract subscriptions from the public to support further research. People are always gullible when there is a prospect of making money without effort — but in this case the speculators' optimism is perhaps not completely irrational. After all, the human body seems to run very well with almost any old fuel and it is not utterly unreasonable to suppose that a machine might be made to work in the same economical way.

In some ways the body does resemble a car engine. Basically it turns the energy of fuel into useful work. It includes systems which can be recognised as a fuel tank, a carburettor, an exhaust pipe and even a supercharger — but the body is vastly more complicated than any man-made engine. In a car, for example, the energy output is related at each moment to the fuel consumption; press on the throttle and more fuel is burnt, with an immediate increase in speed. If we had to take in our food continually we should never get any work done. Consequently the body has to have a mechanism for extracting energy from food, storing it in concentrated but quickly available form and releasing it when needed. A car has a single engine, but the body is driven by hundreds of engines, the muscles, the heart, the lungs, the brain and many others, to each of which energy must be distributed.

If energy was the body's only need, the fuel problem would be difficult enough, but it is immensely complicated by the need for continual replacement of the structure. Skin, worn out blood cells, the linings of the stomach and intestine and many other parts of the body are replaced at short intervals (sometimes lasting for only a few days) and the food handling system must therefore be capable of supplying a wide range of materials for rebuilding. Apart from the obvious replacements, it is now known that every cell and molecule in the body is in a state of constant flux, being torn apart and rebuilt. These rearrangements are carried out in a very economical way, but some energy is inevitably lost and has to be made good from the food intake. The cells of the body need an energy bank, in which energy can be deposited when complex molecules have broken down and withdrawn shortly afterwards for the rebuilding process. All of these operations have to be adaptable to a wide variety of fuels, since man has a more richly varied diet than any other animal.

The problem is complicated because there is only one basic source of energy available to living creatures. Our existence is a continual struggle to capture energy from the sun and to convert it into the highly organised forms needed to sustain life. The same goes for plants and animals — but many of these creatures have a considerable advantage over man in the struggle. Animals (including man) are handicapped by the absence of any process for the direct conversion of sunlight into more useful forms of energy or matter. The only known process of this kind is photosynthesis, the technique by which plants can convert water, carbon dioxide and sunlight into carbohydrates such as sugar and starch. Plants are also the only primary source of proteins, made from simple organic substances combined with nitrogen extracted from the air or soil. A strictly vegetarian diet does not provide all the materials needed to sustain life and we therefore rely on further conversion processes by animals which live on grass and other plants. Though we often think of protein as synonymous with beefsteak, a significant fraction of our daily protein intake comes in bread and potatoes. Asparagus tips are 50% protein, considerably more than is found in steak.

It is possible to run the human engine on a single fuel, by eating nothing but protein. This would however be a costly and, after a while, not a very enjoyable diet. More usually we rely on three main fuels: carbohydrate (such as sugar and starchy foods), fat and protein. The carbohydrates that we eat are without exception the products of plants. Fats on the other hand are mainly derived from animal tissues, with a small contribution from unusual plants such as peanuts and olives. Proteins, as we have seen, come partly from plants at first hand and partly from the flesh and milk of animals which have lived on plants. Carbohydrates (as many people know to their discomfort) are readily converted into fat, but there is no substitute for protein.

The design problems of the body's fuel system may now be summarised. Firstly, it must supply energy for three main purposes: the support of essential organs such as the heart and lungs, the replacement of discarded tissues, and the support of the energy bank needed in the hectic tearing down and building up to which every molecule in the body is constantly subjected. Secondly, the system must be adaptable to a wide range of fuels and must cope with the fact that the intake does not coincide with the demand — indeed the maximum energy output invariably occurs between meals.

Finally, the constituents of the fuel must be broken down into very small fragments before being distributed to the various organs and systems in the body where they are to do their work. This requirement is important for two reasons. Firstly, the important energy exchanges of the body take place only in the cells; consequently the structural spare parts or energy-carrying substances must be reduced to molecules which can pass through cell membranes. There is another and more interesting reason why some of the food intake must be thoroughly broken down. We have already seen (page 117) that all the proteins in the body are personalised, each carrying the individual characteristics of its source, and that the body has a ferocious defence mechanism for protecting and destroying foreign protein. In this situation the protein in food must be reduced to fragments which are too small to have any individuality and therefore suitable for the body to rebuild into its own individual patterns. The problem is that of a Volkswagen dealer faced with a demand for Cadillac spare parts. All that he can do is to break his available spares down into the constituent nuts and bolts, hoping that the new owner will be able to assemble them correctly.

Burning food

Though the requirements imposed on the body's fuel system are exacting, the operating mechanisms are very ingenious and fully capable of dealing with the challenge presented to the designer. Basically the body obtains energy by the combustion of fuel. The combustion is a rather complicated chemical process, not to be confused with the cruder forms of burning — though the energy that the body can extract from a spoonful of sugar is exactly the same as the energy released when the sugar is thrown onto the fire; for this reason the calorie or, more often, the Calorie (equal to a thousand calories) serves as a unit for expressing the potentiality of food or fuel. Many foods are good fuels in both senses; a pound of butter delivers three times as much energy as a pound of TNT.

Much of the body's food intake is eventually reduced to water, carbon dioxide and heat. The chemical energy locked up in food can be reduced to heat with 100% efficiency, but the reverse process can only be accomplished at rather low efficiency. Many of the biochemical processes which occur in the body are designed to catch some of the chemical energy of the food while it is still in a highly organised state and before it has been degraded to heat.

It has already been mentioned that the complex substances present in food must be broken down into simpler materials before the body can use them. A chemist working in a laboratory would find it very difficult to reduce meat or even bread to the simple constituents that the cells of the body can absorb. In setting about this task he would use concentrated acids and high temperatures and would expect some of his experiments to last for several hours. In the body, however, proteins and other complex substances are broken down and put together again very swiftly at a rather low temperature. These successes, which the chemist cannot yet imitate, are brought about by enzymes. An enzyme is essentially a catalyst, that is, a substance which speeds up a chemical reaction and emerges unchanged at the end of the process. Enzymes are all proteins; more than 1,300 different varieties are now known and most cells in the body contain at least two hundred different enzymes. Each enzyme is highly specific for one particular reaction; as the basic biochemical processes are common to all living creatures, many enzymes are found in plants and animals as well as in man.

About two hundred enzymes have been extracted in the laboratory from living matter and purified in crystalline form. The biochemist finds that enzymes can only survive in a test tube if they are kept in the refrigerator, and naturally wonders how they are able to work at body temperature. The answer is that they do not survive very long in the body but are, like the other constituents of living tissue, in a constant state of disintegration and rebuilding.

Digestion

The fuel for a motor car starts its journey in the tank — and the body has a somewhat similar arrangement. The stomach is the tank and the supply pipe is represented by the mouth and gullet. Some preliminary preparation of food occurs in the mouth; chewing helps enjoyment of the flavour, though, contrary to a popular belief, it does little or nothing to help the subsequent digestion. The real work of digestion begins in the stomach which is, among other things, a steriliser, a tenderiser and a liquidiser. The gastric juice, manufactured by cells in the wall of the stomach, contains hydrochloric acid, which kills any germs present in the food and also helps to soften the tougher fibres in meat or other foods. The hydrochloric acid does not, however, play a great part in the initial stages of digestion, which are mainly brought about by pepsin and other enzymes manufactured by cells lining the stomach. The muscle in the stomach wall supports waves of contraction which effectively churn the contents into a fluid mixture. As air and other gases are nearly always trapped in the stomach, the mixing action sometimes produces audible rumbling.

In popular folklore, the acid in the stomach is powerful enough to burn a hole in the carpet. This is a slight exaggeration, but the gastric juice is certainly corrosive. The enzymes which work with it are capable of digesting meat; why do they not digest the wall of the stomach itself? Sometimes of course they do, producing a gastric ulcer which may, in extreme cases, grow into a hole allowing the stomach contents to leak out with serious consequences. In most people, however, the wall of the stomach is adequately protected by the gastric mucosa, a layer of tissue covered by a slippery film of fluid. No convincing explanation is yet available for the immunity of the stomach wall to digestion by its own juices and enzymes. Some substances, particularly aspirin and alcohol, make the mucosa particularly vulnerable and are often associated with stomach bleeding or gastric ulcers. Even in normal people, the gastric lining is probably under a good deal of stress — but the cells of the mucosa are shed at a rapid rate and completely renewed every two or three days. Consequently small patches of damage can be repaired quickly.

Food spends two to four hours in the stomach. A little water and alcohol may be absorbed there, but most of the nutritive content passes on to the next stage of the digestive process as a thick fluid.

Fuel from the petrol tank of a car has to be processed in the carburettor, from which it emerges as a mixture of petrol vapour and air, before useful energy can be extracted from it. In the body, a somewhat similar function is served by the small intestine, though the preparation of the fuel is a good deal more complicated. Some of the food is reduced to simple molecules which can be used to supply energy to the cells and tissues. Other components (such as meat) have to be broken down into small molecules which can then be reassembled to make fresh protein in the characteristic pattern of the individual.

It might, incidentally, be asked why antibodies (page 119) do not go to work immediately and destroy the foreign protein in the stomach. The answer is that the defence against invasion is only activated when foreign protein appears inside the body. The stomach and the rest of the digestive tract are merely a hollow pipe running through the middle and are no more a part of the body than the hole is a part of the doughnut.

Most of the processes of digestion and absorption occur in the small intestine, so called because its diameter (1½–2 inches) is rather less than that of the large intestine which follows it; the length of the small intestine is about twenty feet and it is rather tightly coiled and folded to fit into the abdominal cavity. The first section of the intestine is the duodenum. Its lining receives the acidified contents of the stomach and, not surprisingly, is sometimes eroded with the formation of a duodenal ulcer. Since most of the enzymes involved in digestion will not work in an acid environment, the liquidised food is first neutralised by the injection of fluid from the pancreas, a small organ which lies outside the intestine, close to the duodenum. The pancreatic juice amounts to about a pint and a quarter every day and contains many enzymes; other enzymes are produced in the wall of the intestine and mix with the food as it passes along.

The digestion of fat is rather difficult because most of the intestinal enzymes are soluble only in water. To deal with this problem the liver produces every day nearly a pint of bile, which contains no enzymes but provides bile salts which are essentially detergents. Like the more familiar household washing powders, the bile salts emulsify the oils and fats in the food by breaking them up into tiny drops which mix reasonably well with the watery surroundings; this is very similar to the process by which domestic detergents remove grease from fabrics.

Bile is needed in considerable amounts after every meal. At these times the liver is fully occupied with other tasks, but bile is produced during off-peak periods and stored in concentrated form in the gall-bladder from which it is released when required. The digestion of food in the small intestine is essentially a conveyor belt process with one enzyme after another attacking the product as it passes and reducing the various materials to forms in which they can be absorbed into the tissue linings of the intestine and removed in the circulating blood. Some enzymes convert sugars and starches into glucose. Others convert fats into glycerol (sometimes known as glycerin) and fatty acids, of which acetic acid (the main constituent of vinegar) is a familiar example.

The digestion of proteins is a major task in the small intestine. Apart from what is contributed by food, amounting to 50 to 100 grams per day, a further 150 grams are provided by cells discarded from the lining of the stomach and from the small intestine (which also renews its inner surface every two or three days) as well as from the enzymes in the pancreatic juice. All of these proteins are broken down into amino acids which are the building bricks for the fabrication of proteins. Twenty amino acids are known. A typical protein consists of several hundred amino acid units arranged in a particular order; obviously the number of possible configurations is enormous. When food proteins have been broken down into the basic amino acid units, they have lost their individuality and can be absorbed into the circulating blood without evoking any reaction from the defensive antibodies.

The glucose, glycerol, amino acids and other end products of the digestive process are all absorbed through the wall of the small intestine and eventually reach the large vein which brings them to the liver.

The liver is the body's chemical factory. Though weighing less than four pounds, about 2.5% of the body, it accounts for more than 25% of the body's energy consumption. When a car engine is required to give a performance, it is fitted with a supercharger, which blows extra air into the cylinders and allows the more rapid combustion of fuel. The liver needs a supercharger since its main blood supply comes in veins connected to the small intestine. As we have already seen (page 127) venous blood does not contain much oxygen. The liver has an additional supply in the hepatic artery which delivers well-oxygenated blood at high pressure to mix with the incoming venous blood.

The liver carries out many important functions. It takes glucose, absorbed from the small intestine and redistributes it around the body. Glucose is the major source of energy for the brain and muscle and it is important that its concentration in the circulating blood should be kept reasonably steady. After a meal the glucose level in the blood rises sharply but the surplus is held back by the liver and converted to glycogen, an insoluble starch suitable for storage. The liver can hold enough glycogen to supply the blood for fifteen to eighteen hours; this fills the gap between dinner and breakfast – or, for some people, between dinner and lunch. If the diet is frugal and the glycogen is exhausted, the liver draws on the fat which is stored (sometimes conspicuously) around the body and converts it to glucose. The liver also collects the lactate which is a waste product of muscular exertion and converts it back into glucose which is then available for reuse.

The liver has an important role in dealing with poisonous substances and other materials from which no useful energy or chemical products can be extracted by the digestive system. Some of this work is done by enzymes, but in recent times man's digestive tract has been attacked by a great variety of new drugs, agricultural chemicals and pollutants against which the body has no natural defensive mechanism. Up to a point, the liver can deal with toxic substances by using its available enzymes to promote chemical changes which may make the unwanted materials less dangerous or more soluble and therefore more easily disposed of in the urine. These mechanisms work successfully for poisonous substances produced through the body's normal activities. For example, ammonia (formed in the breakdown of proteins and highly toxic) is converted to urea which, in small amounts at least, is harmless. But when the chemical defences of the liver are challenged by a man-made chemical the outcome may or may not be successful. Barbiturates are converted to less toxic chemicals but tetraethyl lead (found in petrol) and the insecticide parathion, which are non-toxic in themselves, are converted to highly poisonous compounds by enzyme action in the liver.

Digestion and absorption are completed in the small intestine and redistribution of the products is controlled by the liver. The material which reaches the end of the small intestine and passes into the large intestine contains indigestible residues, debris from the walls of the intestine, remnants of bile and pancreatic juice and large numbers of the bacteria which live in the intestine − but very little of nutritive value. The main task of the large intestine is to absorb water for recirculation and to push out what is left.

Most engines discharge some unburnt or partly burnt fuel in their exhaust. The body is quite efficient in this respect, for 99% of the carbohydrate, 92% of the fat and 93% of the protein delivered to the small intestine is absorbed.

The body's fuel system certainly gives an impressive performance, but it has a few shortcomings. Though plants are the basic source of protein, we cannot live on them, because the small intestine is unable to digest the cellulose which forms a large part of the leaves and stems where protein is found. For this reason we have to rely on animals to convert the plant protein into the more digestible form − but at an efficiency of only 10 to 15%. The world's nutritional problems are almost entirely due to protein shortage. If we could eat grass (and obviate the need to use animals as wasteful converters) the world's protein supply would be increased five or ten times. Animals which live largely on herbage have digestive systems with three features lacking in man: a large stomach (60 gallons in the cow), a long intestine (up to 100 feet in the cow) — both allowing for the digestive process to be spread over a considerable period of time — and a population of bacteria able to break down cellulose into starch.

Our diet, varied though it is, has to pass one important test. The twenty amino acids are all needed for the assembly of proteins in the body. Twelve of these twenty are easily dealt with, because any of them can be manufactured if the others are available in sufficient amounts. The remaining eight cannot be made in the body and must therefore be obtained from proteins of other animals. It is for this reason that a completely vegetarian diet will not support life and that substances such as gelatine, though undoubtedly proteins, are of no nutritive value, since they lack many of the essential amino acids. The body is also unable to make most of the vitamins. Vitamin B is in fact produced by some of the bacteria which live in the intestine and vitamin A can be made in the body from carotene, a material found in carrots and other vegetables, but for most of the vitamins we have to rely on the chemical manufacturing processes of other animals.

Remembering, however, that our fuel system allows a wide variety in the choice of food, that it seldom goes wrong — and usually recovers quickly even from the most provoking insults — its performance certainly surpasses that of any chemical plant which could be designed for the same purpose. Not for nothing are the guts associated with determined effort and successful achievement.

11: the control centre

The brain presents baffling problems – to the engineer no less than to the philosopher. Confronted by a new machine or material, the engineer usually manages to learn something about its structure and function by taking a good look at it and by making a few simple tests. Bone is a strong structural material, skin is a tough wrapping and the eye behaves like a camera. No great skill is needed to discover all of this – but the brain is a bigger challenge. Its appearance, a lump of grey jelly, gives no clue at all to what it does. Even when a great many indirect observations have been put together, the brain is curiously difficult to understand or to explain.

The brain has many of the properties of an elaborate computer, but it is immediately obvious that the principles of its design and operation are very different to those commonly used by the electronic engineer. Made of materials that no engineer would ever use, the brain packs an impressive amount of data processing equipment into a small space; the human brain weighs only about three pounds and has a power consumption of about 20 watts. Chunks of the brain can be removed without apparent effect and, in man at least, the brain is capable of abstract thought – an achievement beyond the reach of any computer.

The remarkable abilities of the human brain do not depend on mere size, for whales and elephants have much bigger brains. Even when body weight is taken into account, man is not at the top of the list. Some of the dolphins and porpoises are no bigger than man but have distinctly larger brains; there is of course some evidence that these creatures are rather intelligent and have sophisticated communication systems.

Fossil evidence shows that man's brain has changed little in size and shape during the last 200,000 years — evidence perhaps that the evolutionary process has stopped. Brains are not identical, even in size, but there is no correlation between brain weight and intelligence. The brain of Anatole France weighed under 1,200 grams but Turgenev's was over 2,000 grams. It seems likely that man's advantage in intelligence over other animals is not due to the possession of a bigger brain but to a more complex structure of interconnections between the various parts.

In trying to discover how the brain works, we might first ask how it is made. This, however, is not an easy question to answer because the brain is not a well-defined organ in the same way as the heart or the kidney. It is made of nervous tissue which spreads down the spinal cord and in a network of fine nerves throughout the body. The spinal cord and the rest of the nerves have a task somewhat resembling that of the wiring harness of a car or a complex electronic machine, while the brain itself is responsible for control, navigation, stabilization and other co-ordinating functions.

Since most organs and systems in the body can be modelled in engineering terms, it is natural to ask whether the brain is a computer. It would be more correct to say that it consists of two computers. Anatomists have their own specialised way of describing the structure of the brain, but we shall deal with it in two parts — upper and lower. The lower brain, older in the evolutionary sense, contains the systems essential for the maintenance of life. The upper brain, a relatively recent development, is mainly responsible for memory, emotion, and the refinements of thought and feeling associated with the senses.

The lower part, usually known as the brainstem, grows out of the spinal cord, the thick bundle of nerves which runs through the middle of the backbone. The brainstem is mainly concerned with four tasks. Firstly it controls breathing, blood pressure and a number of other activities which are not under conscious control and which continue whether we are awake or asleep. The cerebellum, a swelling around and behind the brainstem, is the body's automatic pilot. This organ regulates the continual contraction and relaxation of muscles needed to maintain the balance of the body in standing or in motion. The control problems in many simple bodily activities are

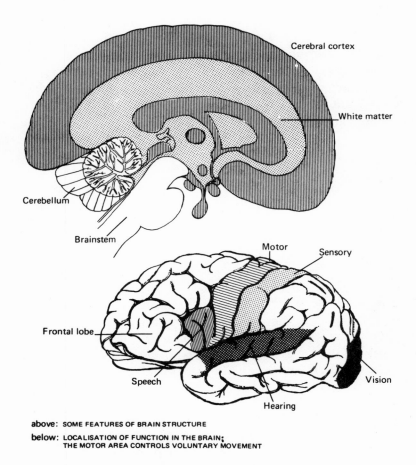

above: SOME FEATURES OF BRAIN STRUCTURE

below: LOCALISATION OF FUNCTION IN THE BRAIN;
THE MOTOR AREA CONTROLS VOLUNTARY MOVEMENT

Figure 10 some features of brain structure (the front of the brain is at the right in the upper drawing and at the left in the lower).

much more complex than those involved in aircraft or missiles, where elaborate computing facilities are provided. It would, for example, be virtually impossible to design a machine to run down stairs with the agility and confidence that the human brain makes possible. Breathing involves the co-ordinated activity of nearly a hundred muscles; raising the hand to the face involves fifty-eight muscles which work on thirty-two bones of the hand and arm.

The brainstem terminates in the old brain, deep inside the skull. This is the seat of the primitive emotions − hunger, fear, pleasure, sex − responsible for the preservation of the individual or the species. Finally the old brain includes some important regulating mechanisms, including the body's thermostat and the so-called appestat, which regulates appetite and food intake.

Brain and mind

In many of the lower animals, the organs that we have just described represent almost the whole of the brain. In man, however, the upper brain has, in the course of evolution, grown very large, giving the human animal many of his distinctive characteristics. The upper brain or cerebrum consists of right and left cerebral hemispheres, partially separated by a deep central fissure. The main part of each hemisphere is the outer layer, a deeply ridged and folded mass of tissue sometimes known as grey matter. This outer layer is the cerebral cortex. The space beneath is occupied by white matter, a solid mass of nerve fibres providing connections between the two hemispheres and other parts of the brain. The cerebral cortex controls speech as well as receiving and processing signals from the various sense organs of the body. The cortex is also the organ mainly concerned with learning, memory, emotions and abstract thought.

We now know a good deal about the organisation of the cortex, largely as the result of direct exploration both in the post mortem room and, more recently, in the living subject.

The idea that different functions were associated with particular locations in the brain was originated by Franz Josef Gall, an Austrian anatomist, at the end of the eighteenth century. He first asserted that the brain was the seat of the mind. Surviving the ecclesiastical censure provoked by this heresy, he went on to suggest that the various functions of the brain are not, as had been supposed, uniformly distributed throughout the cerebral hemispheres but are concentrated in particular places. His ideas grew into the cult of phrenology, which claimed that the strength of various personal qualities could be assessed by feeling the bumps on the outside of the head. History has put Gall among the charlatans, but his basic ideas were correct and were indeed the beginnings of the scientific study of the brain.

The map which was guesswork for Gall now rests on a firmer basis. More than a century ago the British neurologist Hughlings Jackson made important discoveries, starting from the observation that epileptic fits are often preceded by hallucinations of sound, sight, taste, smell or pain. By recording these observations during life and examining the patient's brain after death, he was able to identify the parts of the cortex associated with particular sensations. At about the same time, other experimenters were using tiny electric shocks to stimulate the exposed brains of lightly anaesthetised dogs and were able to identify the parts of the cortex associated with various movements. A big advance was made early in the present century when it became possible to stimulate the brain of the living human subject. The brain itself is not sensitive to pain and, when exposed by the surgeon under local anaesthetic, can be investigated in considerable detail without discomfort to the patient, who remains fully conscious. A small electric current, applied through a needle electrode pushed into the brain, will in one area produce visual images, in another the sensation of sound and in another will cause muscular movement. Sometimes the electrical stimulation will produce a substantial snatch of memory or a dream. These investigations are not motivated merely by curiosity. They are important in the investigation of epilepsy and other disorders; when the affected part of the brain has been located it can be destroyed or removed surgically to cure the disease.

It is remarkable that a procedure as crude as the administration of an electric shock can stimulate the brain to reproduce so many delicate sensations. The brain's insensitivity to pain is perhaps not so remarkable. Most systems of the body are designed in a very economical way and the provision of nerve endings responsive to pain would be a luxury since, covered by its heavy shell of bone, the brain is normally well protected from injury and does not need the early warning system which is useful in other parts of the body.

Small portions of the brain can be removed under local anaesthetic. A surgeon, while engaged in this process, asked the patient: 'What's going through your mind just now?' The patient, spontaneously providing an answer to the problem that has exercised philosophers for centuries, replied: 'A knife'.

This brief look at the outside of the computer gives us some information about what it does but not about how it works. To tackle this problem we need answers to the questions that an electronic engineer would ask about the design of any other computer. A microscopic view of the brain shows something resembling a very complicated wiring diagram, joining a great number of components which are all very similar. These are neurons, of which the brain contains about 10,000 million. Each neuron may be connected to as many as 10,000 others. This multiplicity of cross-connections is important in relation to memory, association and abstract thought.

The neurons or nerve cells are the building bricks of the whole nervous system and, though differing somewhat in size and function, they are all made to a common plan. The body of the cell is a roughly spherical speck of tissue with many tentacles or dendrites. One of these, longer than the rest, is the axon by which impulses generated in one neuron may be passed on to another.

A curious current

The nervous impulse is transmitted by a peculiar kind of electric current. Laboratory experiments on isolated nerve fibres from animals show that the impulse travels along an axon with no loss of strength. This is an achievement which the electrical engineer cannot match, except perhaps in the recently developed superconducting materials which work only at temperatures near absolute zero. In a normal electrical network, a good deal of energy is lost in overcoming the resistance of the conductor. From this point of view, a nerve fibre is a most unlikely material since its electrical resistance is millions of times greater than that of a copper wire. In any event, an ordinary electric current would not flow through a nerve fibre in the living body but would be lost by spreading into the surrounding fluids and tissues, most of which are quite good conductors.

The passage of a nervous impulse along an axon is more like the movement of a spark along a fuse. It depends on a peculiar property, possessed by every living cell. The contents of the cell are held in place by a thin transparent membrane which allows the passage of some substances but not of others. For this reason the composition of the watery part of the cell is not the same as that of the surrounding fluids. In all animal cells, the inside of the membrane has a small negative charge of electricity; axons behave like other cells in this respect.

The nervous impulse is triggered by a localised chemical impulse which alters the permeability of the cell membrane at one point, allowing sodium ions (normally kept out) to pass through. Since each sodium ion carries a positive charge, the negative charge normally maintained within the cell is quickly eliminated. The position is regained by the movement of electrons from the adjoining region of the cell, so that the negative charge is restored at the point where the chemical attack on the cell membrane first occurred. But the adjoining region, having supplied electrons for this purpose, is now itself depleted of negative charge — a process which quickly accelerates because even a small loss of electric charge from the inside of the membrane makes it permeable to positively charged sodium ions from the outside. Consequently this segment of the fibre loses its internal negative charge, the whole process is repeated and the impulse moves along the nerve.

This form of electrical conduction has two important features. The first is that it is an all or nothing process. If the initial chemical insult, reducing the permeability of the membrane, reaches a certain intensity, the impulse starts to travel — but otherwise nothing happens at all. Secondly the impulse, as we have already mentioned, travels quite long distances (two or three feet in some human axons) with no losses. This process is not so mysterious as it appears. The energy required is obtained in chemical form, ultimately from the metabolic processes which maintain the concentration of sodium ions in the surrounding fluid.

The electronic engineer can convey a good deal of information along a wire by varying the strength of the current. In any individual nerve fibre, however, the impulse is always of the same strength. Furthermore the current, if we may call it that, does not flow continually. When one impulse starts, the axon will not accept another until a short time, usually one or two milliseconds, has passed. If the original signal which stimulated the neuron to fire was a large one, impulses will pass along the axon at a high frequency which may reach a few hundred per second; a weak signal will be transmitted at a lower rate. This form of transmission, varying the spacing between successive impulses rather than the strength of the current, is quite well known in the communications industry and has certain advantages, including considerable freedom from external interference — this would certainly be a serious problem in the brain, where large numbers of axons, closely packed together, may be in action at the same time.

Having looked at the connecting wires, we may now return to the general design of the machine. Every computer has the same basic parts: an input device, by which signals or instructions are supplied; a processor, where the arithmetical operations are carried out (in accordance with instructions from a control centre); a store, often called a memory (to hold numbers required or generated in the calculations), and an output device presenting the results of the calculations. The brain uses a great variety of input devices, many of which have already been discussed. These devices are usually transducers, receiving signals in a variety of forms such as heat, light, sound or pressure, and converting them to nervous impulses. These impulses invariably have to pass through many neurons before reaching their final destination.

The nervous system is like a telegraph network with a great many relay stations, receiving messages from one place and passing them on to another. In this process an ingenious chemical system does the job of the operator with his morse key. Contact between one neuron and the next is made in a rather unlikely way. Near its end, the axon divides into a number of branches through which the signal may be transmitted to adjoining neurons by way of the dendrites surrounding the main body of the cell. The axon does not, however, make effective contact with the dendrite but stops short leaving a gap (known as a synapse) about a millionth of an inch across. This situation is not unknown in electrical engineering, where currents often jump small gaps; unfortunately the voltage associated with the nervous impulse is not enough for this purpose.

We have already seen that the triggering of the nervous impulse is brought about by a change in the permeability of the cell membrane, allowing sodium ions to pass through. This change is effected by acetylcholine, a substance which is manufactured on the spot from acetic acid and choline, two materials normally found in cells. As the electrical impulse travels along the axon, a wave of acetylcholine accompanies it. Although the electrical signal cannot jump the synapse, the chemical agent can do so quite easily and this is how the message is transmitted.

Once across the synapse the acetylcholine alters the permeability of the cell membrane in the dendrite and so provokes a fresh nervous impulse which travels until it reaches another synapse, where the process is repeated.

This picture is oversimplified, since it is very unusual for a neuron to be triggered into action by an incoming signal from just one axon. More often, a neuron is receiving signals from a great many axons at the same time and usually several almost simultaneous signals are needed if the chemical effect is to be sufficient to start a fresh impulse from the receiving neuron.

Sometimes an axon is connected, not to another nerve cell but to a muscle. In this situation the outcome is rather similar. The gap between nerve and muscle is bridged by acetylcholine which, on arrival, alters the permeability of the muscle cell membrane and provokes an impulse which causes the muscle fibres to contract.

As the nervous impulse travels down an axon, fresh acetylcholine is produced at each point — but the material already made must be removed when it has served its purpose so that the nerve fibre can return to its normal state in readiness to receive further signals. The material which breaks down the acetylcholine is cholinesterase, which is normally present in the nerves.

Curare (a deadly arrow poison first found in Peru in the sixteenth century) and the nerve gases developed (but not used) during the Second World War are cholinesterase inhibitors; that is to say they block the action of cholinesterase. The effect of introducing one of these substances to the body is to block the transmission of nervous impulses — just as if every Western Union operator in the system suddenly went to sleep. The victim dies because, with his muscles isolated from the control centre, he stops breathing. Synthetic substances related to curare are used in surgery (in small doses) as muscle relaxants, enhancing the effect of anaesthetics.

A versatile switch

The neuron is essentially a switch and it might be wondered whether even 10,000 million switches are enough to make a computer. The answer is that a computer can certainly be made using nothing but switches; however the neuron has a number of other properties which make it more useful than a simple on-off switch. For one thing the neuron, having many attached connections, can do a certain amount of processing on the incoming signals. Some neurons contribute to this process because the chemical released from the ends of their axons is not acetylcholine, but a different substance which makes it harder for the receiving neuron to fire off another impulse. In practice, a receiving neuron is influenced by some signals telling it to fire and by others restraining the trigger. The conflicting signals are added (or subtracted) and, if the final total is above the threshold for firing, another impulse is transmitted – but if not, the message is blocked.

An important factor in the development of electronic computers was the recognition that any mathematical process, however complex, can be broken down into the simplest possible steps. Computers work in binary arithmetic – that is, in the scale of two where the only digits are one and zero and where, in consequence, every arithmetical operation is reduced to a simple yes-no decision. As we have seen, the nervous system works on the same principle. A neuron can transmit an impulse or it can remain silent. Arriving at a synapse, the impulse can cross or can be blocked. In a computer, the choice between stop and go is not random, but is guided by the designer's instructions. In the nervous system the same situation prevails, for the choice between stop and go in each neuron and at each synapse is subject to a variety of built-in influences, allowing the response to be adjusted according to the needs of the situation.

We have seen something of the way in which neurons, the basic elements of the brain, go about their arithmetical processes. We have, in previous chapters, learnt something about the input arrangements and we have also noted some of the output arrangements by which a nervous impulse can be turned into effective action, such as the contraction of a muscle. Other output terminals generate signals regulating the production of hormones (the body's chemical messengers), the heart and respiration rates, and other aspects of the body's activities.

Though it is in many ways a remarkably effective computer, the brain is easily overloaded. It is often difficult to carry on an intelligent conversation while driving a car in city traffic. This is because the information-handling and decision-making capacity of the brain is fully occupied in dealing with the signals which come from the car and the surrounding vehicles. A simple test shows the situation even more clearly. It is quite easy to write the letter A with the right hand and the letter B with the left hand — but most people find it impossible to carry out these simple tasks simultaneously, again because the brain does not have enough capacity for planning the activity and issuing signals to the muscles of the two hands and arms.

The spinal cord

The possibility of overloading would be even more serious but for some very effective means of delegating authority from the brain to lower levels of responsibility in the nervous system. Tread on a drawing pin and the foot will jerk away in about a twentieth of a second; but the time taken for an impulse to travel from the foot to the brain and back again to the appropriate muscles in the leg is about a fifth of a second. For certain actions, including some which may be needed quickly in emergency, the seat of decision is not in the brain but in the spinal cord.

From the brain, twelve pairs of nerves (one on each side)

spread out to the eyes, ears, nose, throat and other parts of the body. Hearing, vision, smell and taste (senses which had an important survival value in primitive man) are controlled directly by the brain. A great many nerves are, however, linked only indirectly with the brain and have their main connections with the spinal cord. Thirty-one pairs of nerves emerge from the spinal cord (passing through openings between the vertebrae) and are then distributed in a rather complicated way, mainly through the trunk and limbs. The spinal cord is linked by various tracts of nervous tissue with the brain.

Touch a hot object and the hand will spring back. It is not until a fraction of a second later that the brain registers the sensation of heat. What happens here is that the heat receptors in the skin cause sensory neurons to fire an impulse which travels to the spinal cord. There the message is passed, possibly through one or two intermediate stages, to motor neurons which send signals to the appropriate muscles causing them to contract and move the hand out of danger. Meanwhile the signals arriving at the spinal cord are being monitored by circuits connected to the brain. The automatic withdrawal of the hand from the heat is an example of a reflex action which happens before the brain has had time to sort out the situation.

The knee-jerk, a test often practiced by physicians, is an example of the simplest kind of reflex involving only two neurons. This reflex depends on the fact that the muscle which forms the upper surface of the thigh terminates in a tendon which is attached to the shin bone. If the subject sits and allows one leg to hang limply over the other, a sharp tap just below the knee cap stretches the tendon and the muscle attached to it. Whenever a muscle is stretched, the nerve endings within it generate a signal which, on arrival at the spinal cord, is immediately re-directed through a motor neuron to the nerve endings which cause the muscle to contract. This stretch reflex is important in maintaining the equilibrium of the body since the stretching of one muscle, if not immediately corrected, would lead to instability. In the situation now under discussion, the contraction of the high muscle exerts a lever action on the lower leg which shoots out. Because the knee-jerk is a reflex involving only two neurons, it is very rapid. It is of no particular importance clinically, but it provides a simple method of checking the integrity of the part of the nervous system including the neurons concerned.

It might be thought that even the reflex arc through the spinal cord is too long for really rapid emergency action. The design of the system is however quite sound. It is important for the record of events occurring at the periphery to be signalled to the brain; this can be done quite well by the links up and down the spinal cord. The position is that of a large organisation in which the head office allows branch managers to decide and act for themselves — but expects to be told what is going on; the system would not work if assistants in different branches were able to establish their own administrative networks independently.

Built-in spare parts

The computer will not work very well if any of its internal connections are cut; a bold slash with the knife will put it out of action completely or reduce its output to gibberish. The brain does rather better. For one thing it is not as vulnerable as it appears. The eyes, ears, lungs and kidneys are duplicated in the body and have a considerable amount of spare capacity so that a useful measure of function can be preserved even if one of the pair is destroyed by disease or accident. The brain, though apparently a single organ, is in fact a pair for many purposes. Each of the cerebral hemispheres can if necessary take over the work of the other. In the normal way the right hemisphere controls many activities on the left-hand side of the body and the left half of the cerebrum looks after the right side of the body. This division of labour is capable of adjustment; corresponding parts of the two hemispheres are linked by bundles of nerve fibres and, in the event of damage, one hemisphere can take over most of the work of the other. The thick bundle of fibres connecting the two hemispheres has been severed in experimental animals and occasionally in human subjects for the treatment of epilepsy. In one patient the two halves of the brain were completely separated by surgery; this drastic procedure led to no serious change in his personality or intelligence. Careful testing did, however, disclose subtler changes. In a right-handed person, the left cerebral hemisphere is dominant. When something is learnt, the memory trace is laid down in the left hemisphere and is available for transfer to the right hemisphere when needed; this is true particularly of memories related to words and language. The patient with his hemispheres separated was not able to read any material in the left half of his visual field or to write with his left hand. More detailed investigations in animals show that when the hemispheres are separated, the animal behaves in many ways as if it had two completely independent brains, each capable of performing a very wide range of mental functions but not sharing their experiences to any useful extent.

Apart from the separation of the hemispheres, the human brain will stand up to other fairly severe mutilations surprisingly well. In 1848, Phineas Gage, the foreman of a road-building team, was injured in a blasting accident when an iron bar went right through his head, causing severe damage to the frontal lobes, the parts of the brain just behind the forehead. He lived for a further twelve years in quite good health, but his personality changed; from being an efficient foreman he became selfish, dreamy and unreliable.

War injuries have given the opportunity for further study of frontal lobe damage, confirming the findings on Phineas Gage. During the 1930s, surgical removal of parts of the frontal lobes was sometimes performed in patients suffering from severe mental depression. In a few cases patients who had been almost unmanageable and in a state where they had little enjoyment from life became placid if not completely apathetic – a result which was sometimes judged to be an improvement. Frontal leucotomy (as the operation is known) is now seldom performed because similar results can be obtained with modern drugs.

The spinal cord is quite well protected by the hollow vertebrae forming the backbone but is vulnerable to accidental damage, for example in road accidents. When the spinal cord is cut, the victim is left with two unconnected nervous systems. Below the cut, the spinal cord is still able to manage its normal reflexes but the parts of the body which it controls have no connection with the brain.

The bladder and bowels are, in very young children, emptied by reflex actions which the more mature human being (or the house-trained pet) learns to control so that evacuation occurs at seemly times and places. Lacking this overriding control from the brain, patients with spinal cord injuries are usually incontinent.

Remembering and forgetting

We have seen something of how the brain organises its work and how it copes with the breakdowns. To complete the picture, we need to know something about the memory. The memory, or information store, of a computer may be held on IBM cards, punched paper tape or magnetic tape. The largest present day computers have access to about as much information as is stored in an average human brain – but man has a great advantage over the machine in rapidity of access to the stored data. To find a particular piece of information by searching through the memory of a computer, held on magnetic tape, may take several seconds. If the computer is being used for calculating a pay roll or a bank balance, this way of working may be quite acceptable. It is possible to make an electronic memory from which data can be retrieved in a fraction of a microsecond – but at quite a high cost. The computer designer therefore usually provides a high speed store of this kind for information and instructions which will be frequently required, but provides the bulk of the memory in a form with longer access times measured in milliseconds or even seconds.

The brain, on the other hand, has uniformly rapid access to the whole information content of the memory, often involving complex searching, association and comparison. The electronic computer does better than the brain in rapid arithmetical calculations but cannot (and will not for a good many years) match the rapid accessibility of information on the scale demonstrated by the brain.

The memory of a computer is a visible block of electronic components and other hardware, but no one knows exactly where the human memory is located. The temporal lobes (lying between the ears and the forehead) are important in this connection since if they are removed by the surgeon, most of the memory is lost. Vivid recollections can be evoked by electrical stimulation of small regions in the temporal lobe; the patient often finds this form of recall much more vivid than the normal memory. However, surgical destruction of the tissues immediately around the tip of the exploring electrodes does not destroy the memory of the events. Much of the information in the memory appears to be distributed throughout a relatively large amount of brain tissue, allowing it to survive localised damage or destruction. This is a trick that the computer designer cannot imitate. Information stored on magnetic tape is precisely localised and is rather vulnerable because, for example, a speck of dust may generate a wrong number with far reaching consequences in all subsequent calculations. Computer programmes usually have built-in checking procedures to detect errors caused by malfunctioning of the machine, but the method used in the brain, with the memory apparently distributed over many millions of storage elements, is a more sophisticated system.

Quite recently, the system known as holography has been developed for the storage of visual images by a somewhat similar technique. A holographic image is, to the naked eye, a meaningless array of dots and smudges − but, when processed by relatively simple optical methods involving lasers and lenses, can reproduce three-dimensional images with excellent detail. A peculiar characteristic of the hologram is that the entire information needed to reconstruct the image is contained in any piece of the plate. If only a small fragment of the original hologram is used, the fine detail is not so clear, but the image is still recognisable. The brain, of course, does not store light signals, but it has been suggested that the electrical changes associated with some aspects of memory may be stored in a system equivalent to a hologram. These speculations are not yet very convincing.

The exact mechanism of memory remains a mystery and, as to location, we know only that the complex aspects of memory, involving recollection and association of visual and auditory experience, appear to reside in the cortex. Some remembering ability is located elsewhere. If the cortex is separated from the rest of the brain in a cat or dog, the animal is still capable of learning simple routine tasks. Since learning involves remembering, the cortex cannot contain the whole of the memory; probably the simpler forms of memory at least are associated with the brain stem.

Simple experiments and observations give a certain amount of information about the process of remembering. After a severe blow to the head, the resulting concussion usually produces temporary loss of memory. When the victim recovers consciousness, he may have no recollection of events which occurred several minutes or even a few hours before the accident. During the succeeding days, the gap is gradually filled but a blank period, which may be a few seconds or a few minutes, remains permanently lost from the memory. Similar results are found after electric shock therapy, in which a large current is passed through the brain.

These observations are linked with the common experience that part of the memory operates on a short term basis. A person's name or telephone number, heard for the first time, can usually be retained in the memory for a few seconds but, if it is not thought to be needed thereafter, it can quickly disappear. The numerous scraps of information that we receive from the outside world are continually filtered, some to be rejected and some to be retained in the long term memory. The process of transfer to the long term store probably takes at least a few minutes and maybe as long as an hour. It seems that information received by the brain stays in a loosely bound state for a few seconds and can be lost by (for example) a severe blow on the head resulting in temporary unconsciousness. Information which has passed through this temporary holding and has been deposited in the long term store is not so easily dislodged and can indeed remain for a lifetime.

How are memory traces laid down in the brain? Not so long ago, it was thought that each fragment of memory was associated with tiny electric currents circulating in a ring of three or more neurons joined in series. But this model does not allow for the fact that memory appears to be quite widely distributed through a considerable mass of brain tissue, and that the memories recalled by electrical stimulation of one cluster of cells in the brain are not lost when the cells are surgically destroyed. More recently the biochemical theory of memory has been vigorously asserted. There is some evidence from animal experiments that the learning process is accompanied by increased production of protein in the brain. The manufacture of protein in the cells of the body is regulated by coded instructions carried on the double helix in the RNA molecule. It has therefore been suggested that memory might be transferred from one animal to another by injection of RNA. Some experimenters have claimed success in this direction, but the theory is not widely supported. RNA taken from one animal and injected into another is inevitably broken down to much simpler forms and it is very doubtful whether any of it reaches the brain of the recipient.

Another interesting speculation starts from the reasonable belief that the neurons, which make up the greater part of the brain tissue, must somehow be involved in memory storage. It has been suggested that the incoming signals modify the synapses which form the connecting links between adjacent neurons, so as to facilitate the passage of subsequent impulses. It is known that all neurons produce occasional spontaneous discharges, without any input signal. The theory suggests that these random impulses will flow along the paths of least resistance – that is, through the synapses already modified by the deposition of memory traces, which will consequently be further strengthened. Like other models of memory, this proposal cannot be proved by experiment but it might explain why the student sometimes finds a difficult problem distinctly easier after he has slept on it.

Though we do not know how to explain memory, it is worthwhile to consider whether it can be improved by drugs or by training. A number of experiments suggests that the rate at which rats learn to perform simple tasks is accelerated by a number of chemicals including strychnine, nicotine and amphetamine (which occurs in purple hearts), but there is no evidence that these or other drugs are effective in man.

Some people have prodigious memories as part of their genetic make up, but there is no convincing evidence that memory can be improved by training or exercise. As the American psychologist William James said nearly a century ago: 'All improvement in memory consists in the improvement of one's habitual methods of recording facts.' In other words, we can make better use of the memory that we have been born with, even if we cannot change it. If we regard the brain as a computer, this conclusion is not really surprising, for it amounts to saying that, though we cannot alter the design of the computer, we may learn how to use it more effectively.

The brain is a most important organ in the body and is by far the most subtle and complex. We can make reasonably good mechanical models, and even mechanical substitutes, for the heart, lung and kidney. We have quite good understanding of the design and function of skin, bone, circulation and respiration. But even the most sophisticated modern technology does not give us the materials or the ideas to make a convincing model of the brain.

12: spare parts

The one-hoss shay applauded by Oliver Wendell Holmes was built in such a logical way that it ran for a hundred years — and then all its parts wore out simultaneously. Our bodies are not so perfectly designed; consequently doctors and engineers spend a lot of effort in trying to repair or replace parts which show signs of failure.

In an ageing population medicine is increasingly concerned with the postponement of death. We all die for the same reason — failure of the oxygen supply to the brain. Indeed the common definition of death nowadays, so important in relation to heart transplants, is the condition in which the electrical activity of the brain has ceased.

Carbon monoxide is poisonous because it prevents haemoglobin from carrying oxygen in the blood. Other poisons, such as cyanide, arsenic and the toxins produced by infectious diseases, work in more complicated ways, by upsetting enzyme reactions or even by destroying red cells, but the eventual result is the same — the brain is suffocated.

Sometimes death comes through the gradual wearing-out of the heart or the kidneys. These are organs with essentially mechanical tasks and it is reasonable to ask whether they can be replaced, like faulty components in any other machine, before a complete breakdown.

At first glance, the prospect seems favourable. Every healthy heart, kidney, lung or liver does its job in the same way and it might seem that spare parts would be available from recently dead people or sometimes from living volunteers; most of us could spare one kidney.

Unfortunately the body is built in a way that seems very strange to the engineer, involving mass-production using components which, though identical by any simple test, are not interchangeable. This incompatibility, causing the body to reject tissue from another person — or indeed any foreign protein, such as an invading microbe — is a valuable defence against infection. Unfortunately the mechanism does not discriminate between dangerous germs and useful replacements.

The rejection mechanism is not infallible (if it was, we should have no infectious diseases) and it is sometimes possible to achieve a compromise, as in blood transfusion. Matching of tissues for organ transplantation is more difficult but the technique is still being developed. The difficulty of replacing faulty organs is magnified because few artificial materials can be left in the body for any length of time without causing trouble. Surgeons and engineers have, however, had some limited success in supplying spare parts.

The most widely used is the cardiac pacemaker, a simple electronic device which has already saved more than 50,000 people from death or permanent invalidism. With one exception, the muscles of the body contract when an electrical signal is received from the brain or elsewhere in the nervous system. The exception is the heart, a complicated muscle with the ability to make rhythmic contractions spontaneously.

In the healthy heart, a wave of contraction spreads from a little group of muscle cells forming the intrinsic pacemaker. This trigger usually fires at sixty to eighty times a minute, though the rate is influenced by signals delivered through nerves in response to emotion, exertion and other factors. The timing impulse normally spreads from the pacemaker region to the ventricles, where most of the heart's pumping effort is generated. If the conducting pathway is blocked, the ventricles contract at their own natural rate, which is only twenty or thirty beats per minute and is not enough to sustain normal life.

The need for an electronic pacemaker is clear — and fortunately the technology is available. A muscle can be stimulated quite crudely, by passing a small electric current through it. The electronic pacemaker comprises a pulse generator (using, in its simplest form, one transistor and a few other components) driven by long-life mercury batteries and delivering a small electrical output to two connecting wires. The pacemaker, enclosed in a plastic shell about the size of a cigarette packet, is usually stitched into some convenient place under the ribs. The output leads can be stitched to the outside of the heart, so that the contraction of the ventricles is stimulated by tiny electric shocks delivered sixty or eighty times a minute — the exact rate can be selected by the surgeon beforehand.

The connecting leads seldom last very long. The heart beats about forty million times a year and the toughest wire usually snaps or comes adrift, necessitating another operation for the patient. A better method is to take the connecting leads (still inside the body) through a convenient vein in the neck and into the heart where they can lie in contact with the muscle of the right ventricle. In this way the stresses on the wires are greatly reduced.

Sometimes the failure of the natural pacemaker is only intermittent and continual stimulation may be dangerous. For such cases, a demand pacemaker is useful. This device monitors the natural electrical activity of the heart and supplies a stimulus only when the natural timing action fails.

Thousands of people lead normal active lives with electronic pacemakers under their ribs — undetectable except by the regular clicks when a transistor radio is brought near!

Though failure of the timing mechanism can be treated by implanting a pacemaker, many people have more serious problems which can be tackled only by replacing the heart itself. Transplants have had little success; the surgical problems are not unduly difficult but the implanted heart is inevitably rejected after a longer or shorter time. While the immunological barrier remains, the prospect of an entirely artificial heart must be explored.

The chances of success do not look good, for several reasons. The natural heart has about the same density as the neighbouring tissues and is securely suspended by large blood vessels and by the pericardium (a tough bag holding the heart) which is anchored to the diaphragm and the gullet. An artificial heart will be a good deal heavier than the natural organ and will probably be made of metal and plastic materials which, in the present state of knowledge, cannot be permanently attached to structures inside the chest cavity. So far we have no man-made materials with which muscle or other tissues will form a union strong enough to withstand the vigorous pulsations of a heart pump.

These problems are still of academic interest because we do not have a suitable power source to drive an artificial heart. Electric motors, nuclear batteries and various mechanical devices all fail through excessive weight and heat production. The gap between nature and artifice is too wide for the most inventive engineer. The heart is built (during embryonic life) by being moulded around the blood. Consequently the flow patterns are all streamlined and the efficiency of the finished machine is very high.

Mechanical hearts, driven by power systems outside the body, have kept calves alive for a few days − but even in experimental animals the totally-implanted artificial heart is not yet a practical possibility. In man, the best that can be done is the heart-lung machine which can, in the operating theatre, take over the control of circulation and respiration for a few hours.

The heart-lung machine has two main parts, a pump to keep the blood moving and an oxygenator to top up the oxygen and allow some of the carbon dioxide to escape. All available pumps have the disadvantage that they damage the blood by destroying red cells. In a healthy person about 1,000 red cells are scrapped every second and an even greater rate can be endured; blackwater fever (a complication of malaria) is so named because the urine becomes darkened with free haemoglobin from damaged blood cells. A patient on the operating table does not have this degree of tolerance and red-cell destruction limits the use of the machine.

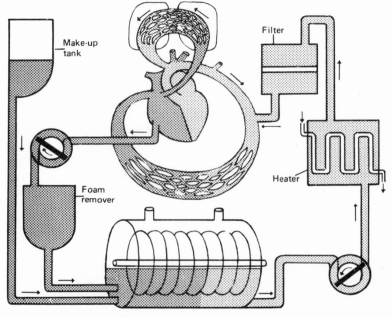

DIAGRAM OF HEART LUNG MACHINE
Some of the blood from the heart is pumped through the oxygenator
(or artificial lung) and returned to the body.

Figure 11 diagram of heart lung machine

The oxygenator is usually a thin plastic film with oxygen on one side and the patient's blood on the other. This part of the machine works reasonably well, though the surface area available for gas exchange does not approach the 1,000 square feet provided in the lungs of a healthy person.

The heart-lung machine is large, crude and costly, requiring a considerable team of expert operators. The prospect of development into a portable device for use at the bedside or in the home seems remote.

The artificial heart is still at a primitive stage – but the artificial kidney is a more satisfactory device, depending on the simple and reliable chemical process of dialysis, in which a plastic membrane allows small molecules (such as salt and glucose) to pass through, but holds back large molecules and blood cells, just as in the natural kidney.

The artificial kidney consists essentially of a pump and a membrane. The patient's blood flows over one side of the membrane and a dialysing fluid flows over the other. Most of the waste products in the blood pass through the membrane — along with most of the useful substances that the body needs. However, if these substances are included in the dialysing fluid (in the same concentrations as in the blood) there is no net loss; if necessary, the concentration of any substance in the blood can be adjusted by altering the amount in the dialysing fluid. Cases of poisoning can often be treated successfully because the toxic materials are washed out of the blood by dialysis.

Until 1960 the artificial kidney (invented by Willem Kolff in 1943) was used mainly to treat poisoning and particularly the kind that occurs when, through kidney failure, too much urea accumulates in the blood. In those days dialysis was a major operation — and it could not be repeated very often. The patient was connected to the machine by two tubes, one in an artery, to take out the blood, and one in a vein, to complete the return path. After the operation the cut artery and vein could not be restored but had to be closed off. Then the blood clotted in the tubes, which could not be used again. There are not many places on the body for fresh connections and patients with chronic kidney failure could not be kept alive.

The solution to this problem was the plastic shunt. Today an artificial kidney patient wears a bandage on one arm. Under it is a short loop of plastic tubing, connecting an artery to a vein. Flowing blood does not clot and the shunt can be opened repeatedly to connect the patient to the machine. The shunt may become infected after two or three years and is then moved to the other arm or to a leg.

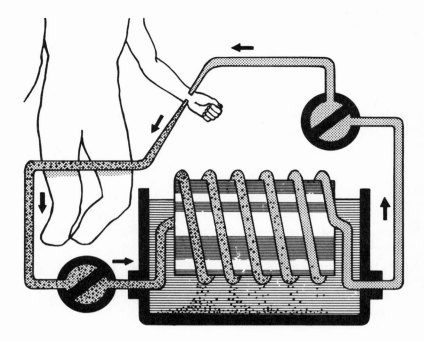

Figure 12 diagrammatic sketch of artificial kidney

Though the artificial kidney is quite large (as big as a washing machine), a single dialysis lasting several hours (during which the patient may sleep) will restore the chemical composition of the blood for a few days' normal life. The present policy is to install kidney machines in the patient's home whenever possible so as to avoid the hazard of cross-infection which has caused many fatalities in hospitals.

The possibility of an artificial kidney comparable in size with its natural counterpart is attractive but well beyond the reach of existing techniques; no dialysing membrane can be made thin enough and strong enough to pack a million dialysing units and thirty miles of piping into a fist-sized machine.

Kidney transplants have become quite successful, with many patients surviving for several years. For reasons not yet understood, the rejection mechanism is less troublesome for a kidney than for a heart.

Kidney failure is a complex process – but simple defects sometimes cause disproportionately serious infirmity. Urinary incontinence or inability to control the emptying of the bladder, is a distressing (and surprisingly common) affliction caused by what is, in engineering terms, merely a leaking tap. The exit of the bladder is normally kept tightly closed by a system of muscles. If these muscles lose their tone (through disease, old age or stresses produced in neighbouring tissues) control of the bladder is lost.

A simple treatment is to supply the appropriate muscles with tiny electric shocks, about twenty times per·second. The resulting contraction keeps the bladder closed. The electrical signals are generated in a radio transmitter (about the size of a packet of cigarettes) which can be carried in the pocket or in a holster under the arm. The transmitter is connected to a plastic-covered aerial coil placed on the skin of the abdomen. A small receiver (needing no batteries) is implanted below the skin and delivers its output by means of two wires terminating in the muscles to be stimulated. The transmitter is normally on; the patient switches it off when he wants to empty his bladder. Modifications of this system have been used successfully in female patients also.

The eye and the ear are such delicate, sensitive and complicated structures that their replacement by transplants or artificial substitutes is beyond the reach of surgery in the foreseeable future. There is however a limited possibility of bypassing a deaf ear or a blind eye. Deafness and blindness usually result from defects in the sense organs themselves and not in the associated regions of the brain. Is it possible to produce the sensations of seeing or hearing by direct electrical stimulation of the brain?

During the 1960s a friendly middle-aged man worked in the library of a medical school in the United States. He wore a hat, even when at his desk, but occasionally took it off, revealing two fifty-pin electrical connectors resting on the top of his head. He was a reprieved murderer, who had taken his freedom in exchange for participation in a bizarre experiment.

Under local anaesthetic, and with X-ray control, one hundred platinum electrodes and half a dozen plastic tubes were inserted at carefully chosen points in his brain. On one day each week, electrical stimuli were applied through the fifty-pin plugs and the responses from other points in his brain were measured with an electro-encephalograph, a sensitive recorder capable of detecting the spontaneous electrical activity of the brain. Sometimes chemical stimuli were applied through the plastic tubes and again the resulting electrical activity was studied.

This co-operative patient reported that the passage of electric current through some parts of his brain gave sensations of sound, though he could not always describe the sounds in terms of normal experience. Work is still in progress and it is certainly possible in principle to deliver messages directly to the auditory cortex, or to the auditory nerve on its way to the cortex.

One problem is to find whether the code by which messages normally pass along the nerve can be imitated by means as crude as an electric current. Another problem is that audible signals are rather complicated, with the information content depending on pitch, presence of harmonics and duration; even a single spoken word is an intricate sound pattern.

For this reason, attention has been concentrated more on the possibility of artificial vision, where a relatively simple black and white pattern could convey useful information to help a blind person. It has been known for half a century that the eye registers a spot of white light whenever the surface of the visual cortex is electrically stimulated; the apparent position of the spot is related to the location of the stimulating electrode.

A system now under development proposes 180 electrodes in a 20 x 9 array, attached to a silicone rubber cap in contact with the brain; the visual cortex is quite easily accessible at the back of the head. The stimulating signals will be generated by electronic circuits fitted in another cap applied to the outside of the skull and linked with a scanning camera. It is hoped that this system will enable a patient to read three-letter groups.

Brain transplants will stay in the realm of science fiction. Other organs can be jerked into activity by electrical insults or by contact with the circulating blood — but the brain is the controller and when it stops it stops for good. The donor must be dead before his brain can be removed, and the definition of death is irreversible cessation of the electrical activity of the brain.

Spare-part surgery is a struggle against three major obstacles. Tissue transplantation, which might be the best answer to most problems, is still impeded by our inadequate understanding of the rejection process. The development of artificial organs is difficult because the body rejects plastics and metal just as it rejects foreign tissues. The intruding material may be pushed out of the body (through the skin) or may be wrapped in fibrous tissue which effectively isolates it. Small passive structures can sometimes be implanted satisfactorily — but not moving systems or devices needing firm attachment to internal organs or tissues. Finally, the sheer complexity of most of the body's mechanisms presents problems that can seldom be solved, even with unlimited access to man-made technology.

13: can man be improved

Can man be improved? It is not difficult to point to weaknesses in design, construction or performance, but some of the shortcomings result from compromises among conflicting requirements, some reflect the unnatural way that we live and some are less important than they appear.

A simple example is provided by the inefficient uptake of oxygen from the lungs. Inspired air contains about 21% of oxygen. The corresponding figure for expired air is 16−17% so that, at the best, only a quarter of the available oxygen is actually taken up. Some of the air that is breathed never reaches the lungs, but lies in the nose and trachea and in the passages leading to the alveoli where oxygen exchange occurs. This air, amounting to about a quarter of each breath, is expired unchanged. The remaining air reaches the alveoli and gives up about a third of its oxygen to the blood. These proportions remain roughly the same during exercise; the increased oxygen requirement of the muscles is then met by deeper and more rapid breathing.

Despite the limited uptake from the breath, the arterial blood is almost saturated with oxygen, because the circulating haemoglobin is loaded to near its limit. Increased transfer of oxygen from the lungs would make little difference to the amount of oxygen carried in the blood.

One of the few examples of inexcusably bad design in the body occurs where it would be least expected, in the coronary circulation which supplies the heart itself. The heart is a muscle and therefore needs an oxygen supply. An engineer might think of taking some of the oxygen from the blood which flows through the heart, but the solution actually found is more complicated. Oxygen is carried to the heart muscle by the two coronary arteries, so called because they encircle the heart like a coronet. They come from the aorta, diverting freshly oxygenated blood from the main supply, which then goes to the other organs and tissues of the body. The first design fault is the lack of standby arrangements to cope with a blockage in one artery or branch. In most parts of the body, neighbouring arteries are joined by sideways connections at intervals so that a reserve circulation is always available for emergencies. But when a patient has a coronary thrombosis, the tissue beyond the blockage in the artery is immediately starved of oxygen. If the patient survives, collateral circulation develops in a day or two — but often the thrombosis proves fatal before then. In other critical organs, such as the brain, a reserve circulation comes into action in a few minutes, but there is no known way of speeding up this process after failure of a coronary artery.

This situation is all the more serious because the coronary arteries are particularly vulnerable to blockage. Most of the arteries in the body become narrower with advancing age — but the coronary vessels are usually the first to suffer complete blockage. The coronary arteries also have to endure more wear and tear than other vessels, since they are compressed at every contraction of the heart. This situation leads to further difficulties because of the resulting intermittent supply of blood, in contrast to the arrangements found elsewhere in the body where most of the major organs have a continuous blood supply.

It seems also that the coronary arteries are too small to do their job properly. In most parts of the body, the arterial blood gives up about a third of its oxygen in passing through the tissues, leaving the remainder to be carried back to the heart in the venous circulation. If the blood supply is reduced for any reason, the percentage extraction of oxygen increases so that the tissues are still adequately supplied. But the blood flowing through the coronary arteries loses about

80% of its oxygen leaving a very small reserve for use in emergency. For all of these reasons, failure of the coronary circulation is a regrettably common cause of disability or death.

The nose, though it performs remarkably well both as an organ of smell and as an air conditioning system, has conspicuous defects. Some of the trouble is caused by the sharp bend in the air passage at the back of each nostril. In most other mammals the lungs are connected to the outside air by a reasonably straight passage – which can be swept out occasionally by sneezing. In man however the sneeze is a very unsatisfactory procedure. Considerable pressure can be generated in the lungs, by the contraction of the diaphragm and the abdominal muscles – but the expired blast cannot negotiate the hairpin bend and most of it escapes through the mouth.

Two legs or four?

The failure of the sneeze is not altogether to be blamed on faulty design. It is partly a consequence of the upright posture which man adopted at a rather late stage in evolution. In changing from four feet to two, the head has to be bent forward to secure a comfortable range of vision. The nose is also appreciably distorted by the great growth of the front part of the brain – which is mainly concerned with thinking and is relatively larger in man than in most other animals.

The upright posture is troublesome in another way. The bones of the face contain four pairs of cavities (called sinuses) lined with mucous membrane. Fluid from the sinuses drains into the nasal cavities and helps to keep them moist. In most animals this drainage is, naturally enough, downhill – but in man the drainage is almost horizontal or even uphill. When the sinuses are infected (as happens in the common cold) the membranes swell, drainage becomes even more difficult and chronic infection may result. Sometimes the narrow exit of a sinus becomes blocked. The blood vessels in the mucous membrane continue to absorb oxygen from the air that remains in the sinus; the consequent reduction in pressure produces the particularly painful vacuum headache associated with sinusitis.

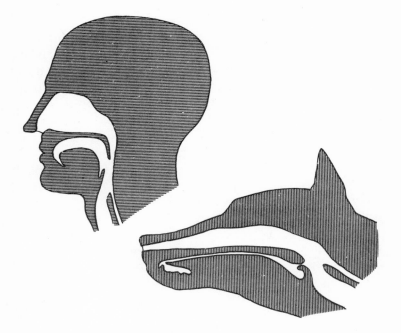

Figure 13 air passages in man and dog

The upright posture is responsible for another painful affliction, the slipped disc. The disc in question here is the intervertebral disc, a structure resembling a squashed golf ball which acts as a shock absorber between the vertebrae in the backbone. The outside of the disc is a springy layer of fibre and cartilage, with a jelly-like material inside. But for the shock absorbers, the jarring of the spine in jumping, climbing stairs or even in walking would be unbearably painful. Sometimes unusual stresses cause a disc to slip out of its seating and to protrude between two vertebrae. The resulting pressure on neighbouring nerves can be very unpleasant.

The occasional inability of the spine to stand up to the wear and tear of life is not the result of faulty design but merely the consequence of the change in the user's requirements. In the original design of the mammalian skeleton, the backbone was a bridge, supported at each end by a pair of limbs. An engineer would not expect to design a bridge and then turn it upright for use as a tower without modification — but this is what man is asking when he walks on two feet. The stress calculations which were correct for the horizontal spine, supported at each end, will not do when the structure is turned into a vertical position. It is not surprising that failures sometimes occur — indeed it is remarkable that the structure adapts so well to the new requirements.

Other parts of the skeleton also suffer from the change. The force on the hip joint may reach four or five times the body weight and the force on the knee joint (in standing up from a squatting position) can be more than half a ton.

There are of course advantages in the upright posture which frees the hands, for carrying food or weapons and for working with tools, three activities which have contributed greatly to the distinctive features of human life as compared with animal life. It is however sometimes thought that more energy is needed to run on two feet than on four. Certainly a man appears to waste energy in accelerating and decelerating the body at the beginning and end of each stride, whereas a four-footed animal moves at a more even rate. The energy cost of running a given distance does not, except for the sprinter, depend much on his speed. A fast runner certainly uses more oxygen, but he covers more ground.

The energy used (in other words the oxygen consumed) in moving one gram of the body over a distance of one kilometre has been measured by experiments on animals of various sizes from the mouse to the horse. The results lie on a smooth curve, large animals being more efficient than small ones, but the results for man are anomalous, indicating an energy expenditure of about twice the value expected for a four-footed animal of the same weight. When these findings were published by Schmidt-Nielsen and his colleagues in the Physiology Department of Duke University, North Carolina, it seemed that the extra energy used in running was the price paid for the advantages of the upright posture. In 1973 however Taylor and Rowntree, at Harvard, trained two chimpanzees and two small monkeys to run on a treadmill either on two or on four legs — and to put up with the equipment needed to measure their oxygen consumption at various speeds. The experimenters found, to their surprise, that the animals used the same amount of oxygen when running on two legs as on four. It is possible therefore that man has secured the advantages of the upright posture without any additional energy cost.

Improving the brain

Sizeable chunks of the brain can be removed with no apparent defect. This prodigality is not extravagant, but reflects the extent to which important mechanisms and connections are duplicated in the interest of reliability. The standby facilities and fail-safe systems occupy so much space that it is reasonable to ask whether the balance between performance and reliability might be altered — in other words whether some of the under-used regions of the brain might be adapted to extend the memory or to provide data processing and control facilities for mental and physical activities at present beyond the capacity of the brain.

For a long time, speculations of this kind were entirely fanciful, in the absence of any method of isolating one part of the brain in order to see, for example, whether it could be stocked with memory traces or programmed in other ways capable of subsequent verification. In 1956 Bures (a Czech physiologist) developed a technique, previously known only as a laboratory curiosity, by which limited regions of the brain can be put out of action for times of up to a few hours. All that is necessary is to apply potassium chloride solution to the exposed brain. In this way Bures, working with rats, was able to put one hemisphere into a dormant state for several hours without affecting the other half of the brain.

In one experiment animals were conditioned (with the help of electric shocks as punishments and food as rewards) to carry out simple tasks such as finding a way through a maze. With one hemisphere inactivated, the rats learned the task without much difficulty. But when this hemisphere was brought to life again and the other half of the brain put out of action, the rats had to learn the task all over again. This meant that memory traces recorded in one hemisphere were not being transferred to the other — as would have occurred had both halves of the brain been in action during the training sessions.

This experiment suggests that the two halves of the brain could be separately programmed, thereby nearly doubling the capacity of the brain.

In the normal way, it is not very difficult to reach the limit of the brain's ability in information handling. Try to write the series 1, 2, 3, . . . with the left hand and at the same time the series 9, 8, 7, . . . with the right hand. Most people find this task exceedingly difficult because, with so many information channels used for exchange of signals, the two halves of the brain (the right half controlling the left hand and the left half the right hand) have limited ability for independent action.

There are of course serious practical difficulties in attempting to separate the two hemispheres. The chemical paralysis which is convenient for experimental animals is clearly not practicable. There would be no point in cutting the connections between the two hemispheres. Even if the two brains are to be separated for programming and control purposes it is desirable for them to have some direct means of communication. Otherwise if the left brain commanded the right hand to make a movement which seemed to the right brain to be undesirable, the only way of correcting the error would be for the left hand to reach out and physically restrain the right hand. It is not impossible that a drug could be found to induce temporary blockage of the bundle of fibres joining the two sides of the brain, so that each hemisphere could be programmed separately and afterwards (with the connections restored) allowed to act independently.

The memory traces concerned with speech (and programming instructions for the many kinds of thinking which have a verbal basis) are usually concentrated in the left hemisphere. It would obviously be inconvenient for both hemispheres to be thinking at the same time and to be giving conflicting instructions to the machinery which produces speech. It is, however, possible that the right hemisphere could, by suitable manipulation of the internal connections in the brain, be pressed into service for language and the associated thinking activities. It might therefore be possible, by temporarily isolating each half of the brain in turn, to make full use of the information processing ability of both hemispheres. Additional control signals would be necessary to make sure that, at any given moment, only one hemisphere was connected to the functional output (for example the hand or the voice), but mechanisms of a similar kind already exist in the brain for other purposes and might therefore be available. The remote prospect (and it must be admitted that there is no early prospect of achieving it) is that the memory or the creativity of man might be significantly extended by making better use of the regions of the brain which are normally quiescent.

Resetting the thermostat

Many machines depend on automatic control systems, though the user expects to be able to reset them if conditions change. The thermostat of a cooker or a central heating plant is easily adjusted, but the human thermostat appears to be tamperproof. Eskimos and Hottentots keep the same body temperature as dwellers in temperate climates. In normal health, the thermostat is never altered; any change in the setting, as in a fever, is a serious sign of illness.

Recent experiments suggest that it may become possible to control the human thermostat. In cats and monkeys the setting can be changed by increasing the concentration of sodium or calcium in the hypothalmus – the region of the brain where the thermostat is located. Sodium raises the temperature and keeps it there, but calcium produces cooling. In these animal studies the thermostat could be set to any temperature between 32 and 42 degrees; the normal level, in man and most other mammals, is about 37 degrees.

The human thermostat is evidently adjusted by the proportion of sodium to calcium in a small region of the brain. Can this ratio be changed without the drastic surgical intervention practised on animals? Certainly a good deal of energy might be saved if the body temperature could be set a little lower in cold weather. The resulting decrease in physical activity and efficiency might, however, be unwelcome. It has been suggested that, with a lower body temperature and correspondingly lower metabolic rate, we should live longer. But the prospect of finding a drug to reset the thermostat seems very remote – and the other consequences of altering the body temperature have not yet been worked out.

The future of bioengineering

What has bioengineering achieved? Progress has so far been notable for large promises and rather small achievements. There are two main reasons for this failure. The first and more important is that bioengineering is not at all like other branches of engineering. In the normal way an engineer, when presented with a problem, has a fairly clear specification of what is required by way of a solution. He can design a machine or a structure, choosing his own materials and constructional methods. Often he can verify the soundness of his design at intermediate stages in construction and he usually has the opportunity at the end of the exercise to see whether his solution meets the client's requirements.

Bioengineering is really engineering in reverse. The starting point is not a clearly defined problem, a repertoire of proven materials and techniques and a blank sheet of paper on the drawing board. The bioengineer starts his work from the wrong end. He is faced with a machine of incomparable subtlety and complexity but does not have the design specification or the operating manual. The machine is built of materials which no conventionally trained engineer would ever think of using. The bones, as we have seen, are made of grit and glue and the rest of the body is largely made out of soup, such as blood and other body fluids, and jelly, such as brain and muscle. Almost every tissue and structural component in the body serves a great variety of different purposes. Electrical signals are conducted through the body in a way which defies Ohm's law, and complicated chemical reactions proceed under conditions of temperature and concentration which a chemical engineer would regard as absurdly inadequate.

The mechanism of the body, in short, avoids or contradicts many basic principles of engineering. Its understanding requires a flexibility of outlook that does not come easily to engineers of traditional upbringing.

A great deal of effort in bioengineering is misdirected through remoteness from the clinical scene. It may be thought that remoteness is no great disadvantage; an aeronautical engineer does not have to spend much of his time as a passenger and a naval architect can do his work quite well on dry land. In these and many other industries, contact between the engineer and the user is not difficult to establish. The situation in bioengineering is very different. Medicine may be a technology but it does not resemble any of the technologies or industries employing engineers.

The first lesson that the bioengineer has to learn is not to underestimate the difficulty of the problems awaiting attention. Rutherford's contemptuous dismissal of biological science, indeed of all science apart from physics, as stamp collecting, may have been made in jest but its echoes can still be heard, encouraging the belief that biological problems are easy or, at any rate, that they can be solved by a brisk application of technical brain power. Bioengineering belongs in the hospital — but too many of today's bioengineers spend their time in university laboratories where they engage in intellectual puzzles remote from the clinical scene. Often they spend their time making computer models of the circulation, the respiration or other systems in the body where the application of ideas derived from control engineering or electronic data processing is superficially attractive. Sometimes they take a problem specified by a doctor and build an instrument to provide the solution. The computer models and the medical equipment made by highly trained and well-meaning engineers remote from the clinical scene sometimes recall the drawing of the aircraft carrier made by a child from a Mexican village. The mast was a totem pole, the aircraft were eagles and the portholes were decorated with Aztec frescoes. The child, transplanted to a completely unfamiliar environment, tried unsuccessfully to bridge the gap by using the only concepts available to him. As Baglivi put it:

Those who apply themselves to several Sciences at the same time are wont to form their Judgments of one by the Precepts and Rules of another.

With a redirection of effort, to bring scientists and engineers nearer to the clinical battlefield, bioengineering can have a spectacular future. Though engineers have been thinking about the mechanism of man for more than three centuries, the contemporary explosion of technology has for the first time given them the power to make positive advances in the understanding and improvement of the human machine.

In modern times, the practice of medicine has been transformed by three revolutions. Sanitation and the control of infection came in the nineteenth century and chemotherapy in the first half of the twentieth century; the bioengineering revolution which may provide the next leap forward is the most exciting and challenging of all the opportunities facing scientists and technologists in the modern world.

Reproduction, the most important of all human activities (since without it the species would disappear) has not yet appeared in our discussion of the mortal machine. The reason for this omission is that biological replication has no mechanical analogue. It is theoretically possible to build a machine which will select components from a store and duplicate itself. But this procedure differs greatly from nature's method. An engineer building an aeroplane does not start with a six-inch model and enlarge it gradually until it flies away. On the contrary, his machine is useless for its intended purpose until the last piece of mechanism has been secured in place.

The process of continual growth and functional development, without disturbing the activities of the existing machine, is completely beyond the scope of engineering, but it is an essential feature of biological replication. In higher animals, including man, the new organism is alive from the moment of conception, that is, from the fusion of egg and sperm, and assumes increasingly complex functions as it grows.

In the kindly sea (only man finds it cruel) the problem of growth is not a serious challenge. Conception is a chancy process, with the female laying thousands of eggs and the male depositing seed in an equally prodigal way, but fertilising only a few of them. In land animals the embryo could not survive the harsh external environment and has to be launched into the world in its own spaceship (egg-shaped) with a food supply enough to bring it to reasonable maturity or, as in man and other mammals, kept under water (the best shock absorber) inside the mother's body and nourished by the liquid conveyor belt connected to the maternal circulation.

Technology can prevent conception and can destroy the embryo before it reaches maturity — but the engineer cannot hope to emulate the subtle, elaborate and (at least in its initial stages) uniquely enjoyable process by which the human machine replicates itself.

index